Embryology at a Glance

Samuel Webster

Lecturer in Anatomy & Embryology
College of Medicine
Swansea University
Swansea, UK

Rhiannon de Wreede

Honorary Lecturer
College of Medicine
Swansea University
Swansea, UK

WILEY-BLACKWELL

A John Wiley & Sons, Ltd., Publication

This edition first published 2012 © 2012 by John Wiley & Sons, Ltd.

Wiley-Blackwell is an imprint of John Wiley & Sons, formed by the merger of Wiley's global Scientific, Technical and Medical business with Blackwell Publishing.

Registered office: John Wiley & Sons, Ltd, The Atrium, Southern Gate, Chichester, West Sussex, PO19 8SQ, UK

Editorial offices: 9600 Garsington Road, Oxford, OX4 2DQ, UK
The Atrium, Southern Gate, Chichester, West Sussex, PO19 8SQ, UK
111 River Street, Hoboken, NJ 07030-5774, USA

For details of our global editorial offices, for customer services and for information about how to apply for permission to reuse the copyright material in this book please see our website at www.wiley.com/wiley-blackwell.

Library of Congress Cataloging-in-Publication Data
Webster, Samuel, 1974-
    Embryology at a glance / Samuel Webster, Rhiannon de Wreede.
        p. ; cm. – (At a glance series)
    Includes bibliographical references and index.
    ISBN 978-0-470-65453-8 (pbk. : alk. paper)
    I. De Wreede, Rhiannon. II. Title. III. Series: At a glance series (Oxford, England).
    [DNLM: 1. Embryonic Development. QS 604]
    612.6'4–dc23
                                    2011049102

A catalogue record for this book is available from the British Library.

Wiley also publishes its books in a variety of electronic formats. Some content that appears in print may not be available in electronic books.

Cover image: © Joseph Mercier | Dreamstime.com
Cover design by Meaden Creative

Set in 9/11.5pt Times by Toppan Best-set Premedia Limited
Printed and bound in Malaysia by Vivar Printing Sdn Bhd

1   2012

Contents

Companion website

This book is accompanied by a website containing a link to Dr Webster's website and podcasts:

www.wiley.com/go/embryology

Preface

We wrote this book for our students; those studying medicine with us, those listening to the podcasts wherever they may be, and those studying the other forms that biology takes on their paths to whatever goals they may have in life. We have introduced many students to the fascinating and often surprising processes of embryological development, and we hope to do the same in this book. It is written for anyone wondering, "where did I come from?"

The content of this book extends beyond the curricula of most medicine, health and bioscience teaching programmes in terms of breadth, but we have limited its depth. Many embryology text-books cover development in detail, but students struggle to get started, and to get to grips with early concepts. Hopefully we have addressed these difficulties with this book.

We hope that you will use this book to begin your studies of embryology and development, but also that you will return to it when preparing for assessments or checking your understanding. You will find example assessment questions in Chapters 46 and 47, and a glossary in Chapter 48.

Let this be the start of your integration of embryonic development with anatomy, to the ends of improved understanding and better patient care or scientific insight.

Acknowledgements

Thank you to Kim and Robin for being so encouraging and putting up with the time demands of completing this book. We would also like to thank the editors at Wiley-Blackwell for leading us through this process and for their support and encouragement, and Jane Fallows for all her work with the illustrations.

List of abbreviations

AER	Apical ectodermal ridge	**IVC**	Inferior vena cava
CAM	Cell adhesion molecule	**IVD**	Intervertebral disc
CN	Cranial nerve	**IVF**	*In vitro* fertilisation
CSF	Cerebrospinal fluid	**LH**	Luteinising hormone
ECMO	Extracorporeal membrane oxygenation	**LMP**	Last menstrual period
FGF	Fibroblast growth factor	**PDA**	Patent ductus arteriosus
FSH	Follicle stimulating hormone	**PFO**	Patent foramen ovale
GnRH	Gonadotrophin releasing hormone	**PTH**	Parathyroid hormone
HbF	Foetal haemoglobin	**PZ**	Proliferating zone
hCG	Human chorionic gonadotrophin	**Rh**	Rhesus
hCS	Human chorionic somatomammotrophin	**SVC**	Superior vena cava
IUD	Intrauterine device – contraceptive device	**TGF**	Transforming growth factor
IUGR	Intrauterine growth restriction	**ZPA**	Zone of polarising activity

Timeline

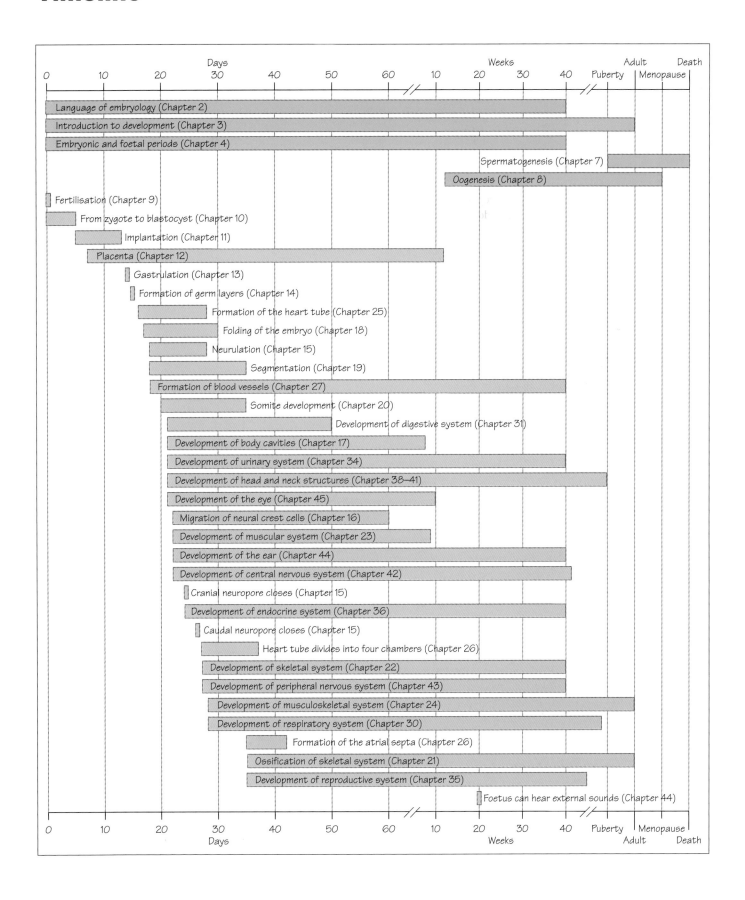

1 Embryology in medicine

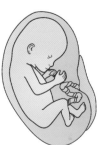

Figure 1.1
The early embryo develops from a simple group of cells into complex shapes and structures in the early weeks

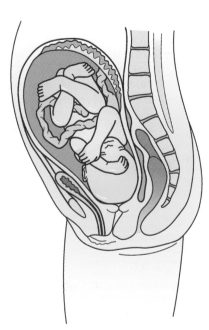

Figure 1.2
Development continues beyond embryology and the foetus continues to grow and mature

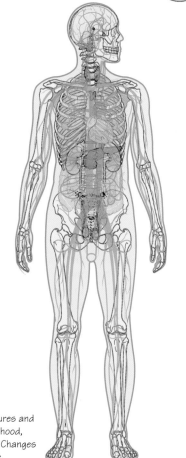

Figure 1.3
Development of biological structures and systems continues through childhood, adolescence and into adulthood. Changes continue to occur throughout life

Embryology at a Glance, First Edition. Samuel Webster and Rhiannon de Wreede.

What is embryology?

Animals begin life as a single cell. That cell must produce new cells and form increasingly complex structures in an organised and controlled manner to reliably and successfully build a new organism (Figures 1.1 and 1.2). As an adult human may be made up of around 100 trillion cells this must be an impressively well-choreographed compendium of processes.

Embryology is the branch of biology that studies the early formation and development of these organisms. Embryology begins with fertilisation, and we have included the processes that lead to fertilisation in this text. The human embryonic period is completed by week 8, but we follow development of many systems through the foetal stages, birth and, in some cases, describe how changes continue to occur into infancy, adolescence and adult life (Figure 1.3).

Aims and format

This book aims to be concise but readable. We have provided a page of text accompanied by a page of illustrations in each chapter. Be aware that the concise manner of the text means that the topic is not necessarily comprehensive. We aim to be clear in our descriptions and explanations but this book should prepare you to move on to more comprehensive and detailed texts and sources.

Why study embryology?

Our biological development is a fascinating subject deserving study for interest's sake alone. An understanding of embryological development also helps us answer questions about our adult anatomy, why congenital abnormalities sometimes occur and gives us insights into where we come from. In medicine the importance of an understanding of normal development quickly becomes clear as a student begins to make the same links between embryology, anatomy, physiology and neonatal medicine.

The study of embryology has been documented as far back as the sixth century BC when the chicken egg was noted as a perfect way of studying development. Aristotle (384–322 BC) compared preformationism and epigenetic theories of development. Do animals begin in a preformed way, merely becoming larger, or do they form from something much simpler, developing the structures and systems of the adult in time? From studies of chickens' eggs of different days of incubation and comparisons with the embryos of other animals Aristotle favoured epigenetic theory, noting similarities between the embryos of humans and other animals in very early stages. In a chicken's egg, a beating heart can be observed with the naked eye before much else of the chicken has formed.

Aristotle's views directed the field of embryology until the invention of the light microscope in the late 1500s. From then onwards embryology as a field of study was developed.

A common problem that students face when studying embryology is the apparent complexity of the topic. Cells change names, the vocabulary seems vast, shapes form, are named and renamed, and not only are there structures to be concerned with but also the changes to those structures with time. In anatomy, structures acquire new names as they move to a new place or pass another structure (e.g. the external iliac artery passes deep to the inguinal ligament and becomes the femoral artery). In embryology, cells acquire new names when they differentiate to become more specialised or group together in a new place; structures have new names when they move, change shape or new structures form around them. With time and study students discover these processes, just as they discover anatomical structures.

Embryology in modern medicine

If a student can build a good understanding of embryological and foetal development they will have a foundation for a better understanding of anatomy, physiology and developmental anomalies. For a medical student it is not difficult to see why these subjects are essential. If a baby is born with 'a hole in the heart', what does this mean? Is there just one kind of hole? Or more than one? Where is the hole? What are the physiological implications? How would you repair this? If that part of the heart did not form properly what else might have not formed properly? How can you explain to the parents why this happened, and what the implications are for the baby and future children? A knowledge of the timings at which organs and structures develop is also important in determining periods of susceptibility for the developing embryo to environmental factors and teratogens.

Why read this book?

We appreciate that the subject of embryology still induces concern and despair in students. However, if it helps you in your profession you should want to dig deep into the wealth of understanding it can give you. We also appreciate that you have enough to learn already and so this book hopes to represent embryology in an accessible format, as our podcasts try to do.

One thing that has not changed with the development of embryology as a subject is that the more information that is gathered, the more numerous are the questions left unanswered. For example, we barely mention the molecular aspects of development here. Should your interest in embryology and mechanisms of development be aroused by this book, we hope that you will seek out more detailed sources of information to consolidate your learning.

Language of embryology

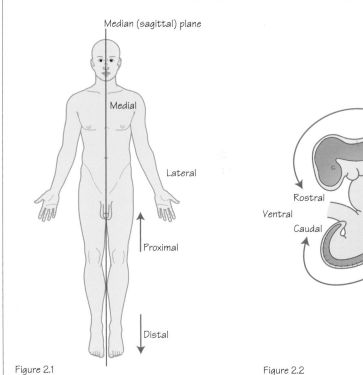

Figure 2.1

The anatomical position

The adult anatomical position can be used to describe structures that are medial or lateral relative to the median sagittal plane, and proximal or distal in the limbs. These also apply to the embryo

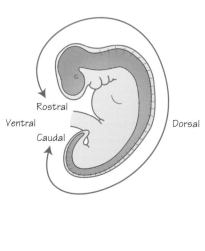

Figure 2.2

The surfaces of the embryo that rostral, caudal, dorsal and ventral refer to

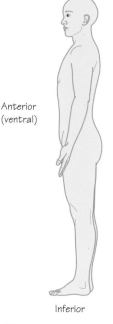

Figure 2.3

Note how the descriptions of superior, inferior, anterior and posterior of the adult anatomical position relate to the descriptions of the embryo

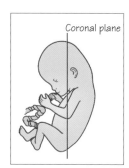

Figure 2.4

The coronal plane in the embryo and the adult refer to a plane of section cut like this

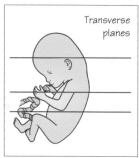

Figure 2.5

Transverse planes are cut across the embryo as in this diagram, perpendicular to the coronal plane

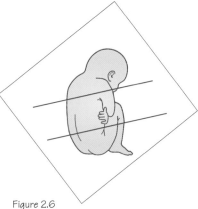

Figure 2.6

Oblique planes are cut in directions unlike the other planes. They do not cut along a clear X, Y or Z axis

Embryology at a Glance, First Edition. Samuel Webster and Rhiannon de Wreede.
12

Time period: day 0–266

Introduction

The language used to describe the embryo and the developmental processes that mould it is necessarily descriptive. It is similar to anatomical terminology, but there are some common differences that the reader should be aware of.

The embryo does not, and for most of its existence cannot, take on the anatomical position. The embryo is more curved and folded than the erect adult. The adult anatomical position is described as the body being erect with the arms at the sides, palms forward and thumbs away from the body (Figure 2.1). The anatomical relationships of structures are described as if in this position, so for the embryo we need to rethink this a little.

Cranial–caudal

Anatomically speaking, you may interchangeably use cranial or superior, and caudal or inferior. Cranial clearly refers to the head end of the embryo and caudal (from the Latin word *cauda*, meaning 'tail') refers to the tail end (Figure 2.2). If you imagine the early sheet of the embryo with the primitive streak (see Chapter 13) showing us the cranial and caudal ends, you can imagine that it can be clearer to use these terms rather than superior and inferior.

The term 'rostral' may also be used in place of cranial. Rostral is derived from the Latin word *rostrum*, meaning 'beak'.

Dorsal–ventral

The dorsal surface of the embryo and the adult is the back (Figure 2.2). Dorsal also refers to the surface of the foot opposite to the plantar surface, the surface of the tongue covered with papillae, and the superior surface of the brain, so some care is needed.

The ventral surface of the embryo is the front or anterior of the embryo, opposite the dorsal surface.

Medial–lateral

As with adult anatomy, structures nearer to the midline sagittal plane are more medial, and structures further from the midline are more lateral (Figure 2.3). This also helps us describe the left–right axis of the embryo.

Proximal–distal

Proximal and distal are a little different from medial and lateral, but similarly describe structures near to the centre of the body (proximal) and further from the centre (distal) (Figure 2.1). These terms are typically used to describe limb structures. The hand is distal to the elbow, for example.

Sections

Often, to show the parts of the embryo being described, illustrations must be of a section of the embryo or a structure. These sections may be transverse, median, coronal or oblique. You can see these planes of sections in the illustrations on the opposite page (Figures 2.4–2.6).

3 Introduction to development

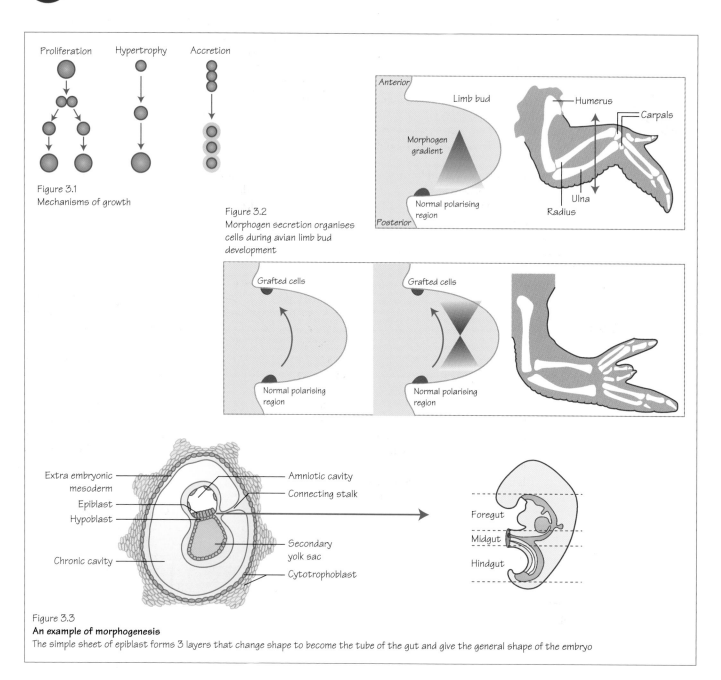

Figure 3.1
Mechanisms of growth

Figure 3.2
Morphogen secretion organises cells during avian limb bud development

Figure 3.3
An example of morphogenesis
The simple sheet of epiblast forms 3 layers that change shape to become the tube of the gut and give the general shape of the embryo

Time period: day 0 to adult

Development

Development, in this book, describes our journey from a single cell to a complex multicellular organism. Development does not end at birth, but continues with childhood and puberty to early adulthood.

We must describe how a cell from the father and a cell from the mother combine to form a new genetic individual, and how this new cell forms others, how they become organised to form new shapes, specialised interlinked structures, and grow. With this knowledge we become able to understand how these processes can be interfered with, and how abnormalities arise.

Growth

Growth may be described as the process of increasing in physical size, or as development from a lower or simpler form to a higher or more complex form.

In embryology, growth with respect to a change in size may occur through an increase in cell number, an increase in cell size or an increase in extracellular material (Figure 3.1).

Embryology at a Glance, First Edition. Samuel Webster and Rhiannon de Wreede.

Increasing cell number occurs by cells dividing to produce daughter cells by **proliferation**. Proliferation is a core mechanism of increasing the size of a tissue or organism, and is also found in adult tissues in repair or where there is an expected continual loss of cells such as in the skin or gastrointestinal tract. Stem cells are particularly good at proliferating.

An increase in cell size occurs by **hypertrophy**. In adults, muscle cells respond to weight training by hypertrophy, and this is one way in which muscles become larger. During development, hypertrophy of cartilage cells during endochondral ossification is an important part of the growth of long bones. Be aware that the term hypertrophy can also be used to describe a structure that is larger than normal.

Cells may surround themselves with an extracellular matrix, particularly in connective tissues such as bone and cartilage. By **accretion** these cells increase the size of the tissue by increasing the amount of extracellular matrix, either as part of development or in response to mechanical loading.

Cells may also die by programmed cell death, or apoptosis. This might be considered an opposite to growth, and in development is an important method of forming certain structures like the fingers and toes.

Differentiation

During development, cells become specialised as they move from a multipotent stem cell type towards a cell type with a particular task, such as a muscle cell, a bone cell, a neuron or an epithelial cell. When the cell becomes more specialised it is considered to have **differentiated** into a mature cell type. If that cell divides, its daughter cells will also be of that mature cell type.

In humans, a mature cell is unlikely to dedifferentiate back into a stem cell, but the process by which this can occur is being exploited in the laboratory with the aim of producing stem cells from adult tissues. These stem cells could then be pushed to differentiate into the cell type needed to grow new tissue or treat a disease.

Signalling

A signal from one group of cells influences the development of another (adjacent, nearby or distant) group of cells. Hormones act as signals, for example. For a cell to be affected by a signal it must possess an appropriate receptor.

In the embryo the signalling of a vast array of different proteins by different groups of cells allows those cells to gain information about their current and future tasks, be that migration, proliferation, differentiation or something else.

Organisation

Early in development the ball of cells or simple sheets of the embryo do not give much clue about which cells will form which structures. It is difficult to determine which part will become the head and which will become the tail. However, the cells are aware of their position and the roles that they will have and we can see this by looking at the signalling proteins and connections between cells.

For example, the upper limb begins to develop as a simple bud of cells. The cells in that bud must be organised to produce the structures of the arm, the forearm and the hand. The ulna bone must form in the right place relative to the radius, and the thumb must form appropriately in relation to the fingers. This may occur partly because a group of cells on the caudal aspect of the limb bud produces a morphogen that diffuses across the early limb bud (Figure 3.2). Cells near the site of morphogen production experience a high concentration, and cells further away on the cranial side of the bud experience a lower concentration. Development of these cells progresses differently as a result. If experimentally you transplant some of the morphogen-producing cells to the cranial part of the limb bud, duplicate digital structures form. See Chapter 23 for more about limb development.

This is one example of how cells organise themselves and others during development. With organisation, structure follows.

Morphogenesis

The formation of shape during development is **morphogenesis**. Cells are able to change the ways in which they adhere to one another, they can extend processes and pull themselves along, migrating to new locations, and they can change their own shapes. In a tissue there may be a change in cell number, cell size or accretion of extracellular material. In these ways a tissue gains and changes shape.

An early example of morphogenesis in embryonic development occurs with the change from simple flat sheets of cells to the rolled up tubes of the embryo and gastrointestinal tract (Figure 3.3). A simple structure has become more complex. Chapter 13 covers this in more detail.

Clinical relevance

Interruptions of signalling, proliferation, differentiation, migration, and so on, cause congenital abnormalities. **Teratogens** that affect development during key periods may have significant effects. For example, if the drug thalidomide is taken during early limb development it can cause phocomelia (hands and feet attached to abnormally shortened limbs). Other environmental factors and genetic mutations can cause abnormal development. The embryo is most sensitive during weeks 3–8.

Dysmorphogenesis is a term used for the abnormal development of body structures. It may occur because of malformation or deformation. If the processes required to normally form a structure fail to occur the result is a malformation. If the neural tube fails to close, for example, the resulting neural tube defect is a malformation. A deformation occurs if external mechanical forces affect development. For example, damage to the amniotic sac can cause amniotic bands that may wrap around developing limbs and cause amputation of limbs or digits.

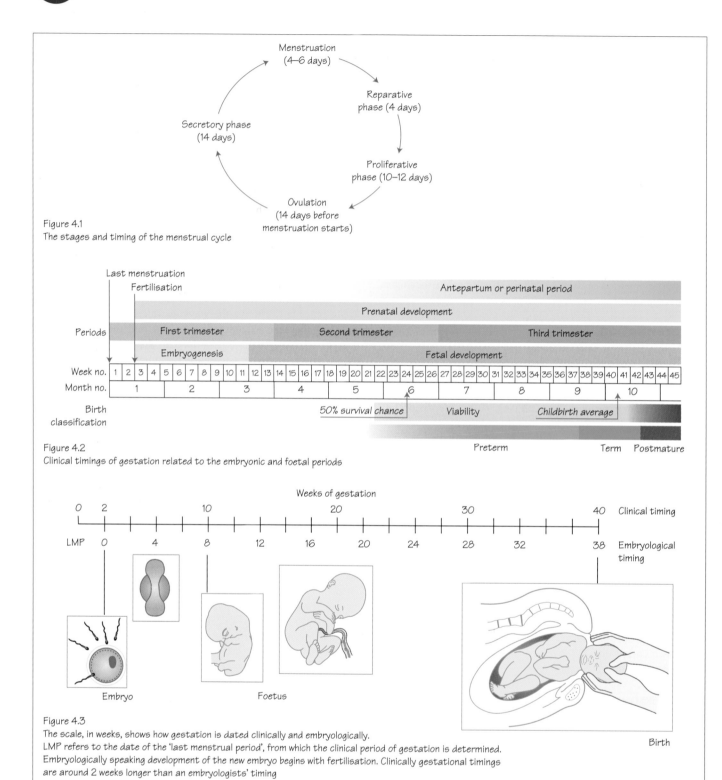

Figure 4.1
The stages and timing of the menstrual cycle

Figure 4.2
Clinical timings of gestation related to the embryonic and foetal periods

Figure 4.3
The scale, in weeks, shows how gestation is dated clinically and embryologically.
LMP refers to the date of the 'last menstrual period', from which the clinical period of gestation is determined.
Embryologically speaking development of the new embryo begins with fertilisation. Clinically gestational timings are around 2 weeks longer than an embryologists' timing

Time period: day 0 to birth

Embryonic period

The embryonic period is considered to be the period from fertilisation to the end of the eighth week. The period from fertilisation to implantation of the blastocyst into the uterus (2 weeks) is sometimes called the period of the egg.

During the period of the egg the early zygote rapidly proliferates to produce a ball of cells that makes its way along the uterine tube towards the uterus. The complexity of the blastocyst increases as it progresses towards the site of implantation.

During the embryonic period the major structures of the embryo are formed, and by 8 weeks most organs and systems are established and functioning to some extent, but many are at an immature stage of development. At the end of the eighth week the external features of the embryo are recognisable; the eyes, ears and mouth are visible, the fingers and toes are formed, and limbs have elbow and knee joints.

Foetal period

From the ninth week to birth the foetus matures during the foetal period. The foetus grows rapidly in size, mass and complexity, and its proportions change (for example, head to trunk, and limbs). The foetus' weight increases considerably in the latter stages of the foetal period. Organs and systems continue in their functional development, and some systems change considerably at birth (for example, the respiratory and circulatory systems).

Birth in humans normally occurs between 37 and 42 weeks after fertilisation.

Trimesters

The nine calendar month gestation period is split into 3-month periods called trimesters. During the first trimester the embryonic and early foetal periods occur. In the second trimester the uterus becomes much larger as the foetus grows considerably, and symptoms of morning sickness tend to subside. A foetus in the third trimester turns and the head drops into the pelvic cavity (engagement) in preparation for birth. Babies born prematurely during the third trimester may survive, particularly with specialised intensive care treatment.

Clinical and embryological timings

Embryologists use timings from the date of fertilisation, and all the timings in this book will relate to that time. Embryologists studying the embryos of animals often have an advantage in being able to fairly accurately note when fertilisation occurred. Clinically, the date of fertilisation is more difficult to determine.

A woman's menstrual cycle will take around 28 days to complete, starting with the first day of the menstrual period (bleed) and returning to the same point (Figure 4.1). Menstruation occurs for 3–6 days, followed by the proliferative phase for 10–12 days. Ovulation occurs around 14 days before the start of the next menstrual period. If the released ovum is fertilised menstruation will not occur. Fertilisation must occur within 1 day of ovulation.

The event of the last menstrual period can be used to date the period of gestation clinically, although the date on which fertilisation took place will be uncertain because of variability in the length of the cycle between the start of menstruation and ovulation.

Clinically, gestational timings are around 2 weeks longer than an embryologist's timing (Figure 4.2). If the embryonic period is complete at the end of week 8, a clinician would record this as the end of week 10 (Figure 4.3).

Clinical relevance

If you are a medical, nursing or health sciences student then you must be aware of the 2-week difference between embryologists' and clinicans' gestation timings.

A gestation period of 40 weeks is equal to 10 lunar months. A period of 10 lunar months is, on average, 7 days longer than any 9 calendar months. Using the mother's date of the start of her last menstrual period you can quickly calculate an estimated date of delivery by adding 9 calendar months and 7 days.

An awareness of the period of the egg, the embryonic period and the trimesters helps understand the periods of susceptibility of the embryo and the foetus. For example, after the period of the egg and during the embryonic period the embryo is particularly vulnerable to the effects of teratogens and environmental insults. The respiratory system develops significantly during the third trimester, so linking the timing of a premature birth to the potential requirements of the baby are important.

5 Mitosis

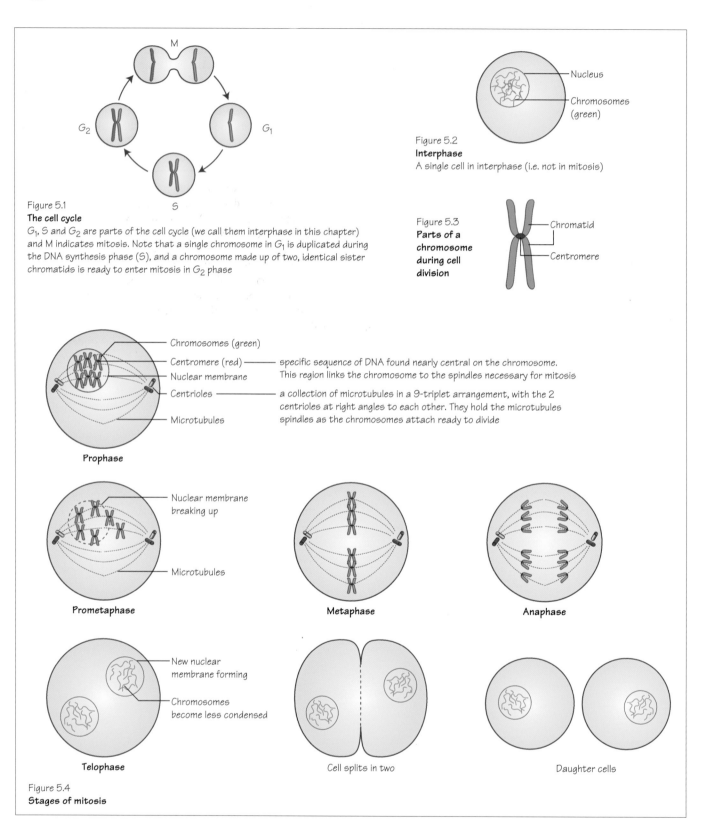

Figure 5.1
The cell cycle
G_1, S and G_2 are parts of the cell cycle (we call them interphase in this chapter) and M indicates mitosis. Note that a single chromosome in G_1 is duplicated during the DNA synthesis phase (S), and a chromosome made up of two, identical sister chromatids is ready to enter mitosis in G_2 phase

Figure 5.2
Interphase
A single cell in interphase (i.e. not in mitosis)

Nucleus
Chromosomes (green)

Figure 5.3
Parts of a chromosome during cell division

Chromatid
Centromere

Prophase

Chromosomes (green)
Centromere (red) — specific sequence of DNA found nearly central on the chromosome. This region links the chromosome to the spindles necessary for mitosis
Nuclear membrane
Centrioles — a collection of microtubules in a 9-triplet arrangement, with the 2 centrioles at right angles to each other. They hold the microtubules spindles as the chromosomes attach ready to divide
Microtubules

Prometaphase

Nuclear membrane breaking up
Microtubules

Metaphase

Anaphase

Telophase

New nuclear membrane forming
Chromosomes become less condensed

Cell splits in two

Daughter cells

Figure 5.4
Stages of mitosis

Embryology at a Glance, First Edition. Samuel Webster and Rhiannon de Wreede.

Time period: day 0 to adult

Cell division

Cell division normally occurs in eukaryotic organisms through the process of **mitosis**, in which the maternal cell divides to form two genetically identical daughter cells (Figure 5.1). This allows growth, repair, replacement of lost cells and so on. A key process during mitosis is the duplication of DNA to give two identical sets of chromosomes, which are then pulled apart and new cells are formed around each set. The new cells may be considered to be clones of the maternal cell.

Mitosis

A cell dividing by mitosis passes through six phases.
- **Interphase:** the cell goes about its normal, daily business (Figure 5.2). This is also known as the cell cycle, and includes phases of its own: G_1 (gap 1), S (synthesis) and G_2 (gap 2). DNA is duplicated (synthesised) during S phase.
- **Prophase:** DNA condenses to become chromosomes which are visible under a microscope (Figure 5.3). Centrioles move to opposite ends of the cell and extend microtubules out (this is the mitotic spindle). The centromeres at the centre of the chromosomes also begin to extend fibres outwards (Figure 5.4).
- **Prometaphase:** the nuclear membrane disappears, microtubules attach centrioles to centromeres and start pulling the chromosomes.
- **Metaphase:** chromosomes become aligned in the middle of the cell.
- **Anaphase:** chromosome pairs split (centromeres are cut), and one of each pair (sister chromatids) move to either end of the cell.
- **Telophase:** sister chromatids reach opposite ends of the cell and become less condensed and no longer visible; new membranes form around the new nuclei for the daughter cells.
- **Cytokinesis:** an actin ring around the centre of the cell shrinks and splits the cell in two.
- **Interphase:** the cell goes about its normal, daily business (including preparing for and doubling its DNA to form pairs of chromosomes).

Clinical relevance

Errors in mitotic division, although rare, will be carried into the daughter cells of that division, and onwards to new cells produced from them. Errors in early embryonic development could have catastrophic consequences, as an error in one cell would quickly become an error in a huge number of cells. Chromosomal damage can give small or significant effects, such as trisomy (an extra copy of a chromosome), or translocation or inversion of a broken section. Trisomy 21, for example, results in Down syndrome.

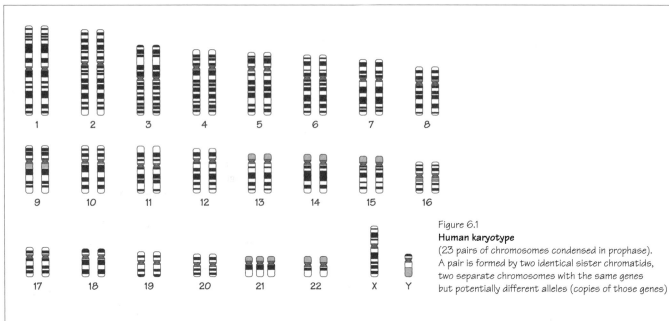

Figure 6.1
Human karyotype
(23 pairs of chromosomes condensed in prophase).
A pair is formed by two identical sister chromatids,
two separate chromosomes with the same genes
but potentially different alleles (copies of those genes)

Figure 6.2
A chromosome in the G_1 phase
after mitosis (interphase)

Figure 6.3
A chromosome after S phase. The DNA has been
duplicated to produce two identical sister chromatids

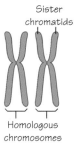

Sister
chromatids

Homologous
chromosomes

Figure 6.4
A homologous pair of chromosomes (meiosis)

Red and green strands are pairs of (homologous) chromosomes
(a pair has one red and one green chromosome). The red strand
signifies the paternal DNA and the green strand the maternal
DNA within this cell

Homologous chromosomes begin to
swap sections of DNA (alleles)

| Interphase | Prophase I | Prometaphase I | Metaphase I | Anaphase I | Telophase I | Interphase II |

Figure 6.5
Meiosis I is similar to mitosis, but at the end of meiosis I two cells have formed, each with one chromosome of a homologous pair.
They are haploid cells. Note the crossover of alleles between homologous pairs

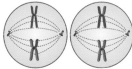 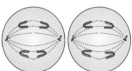 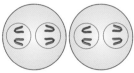 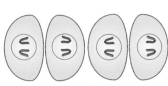

| Metaphase II | Anaphase II | Telophase II | 4 haploid cells |

Figure 6.6
In the two haploid cells division begins again. At the end of meiosis II four haploid cells have formed, each with 23 chromosomes
(not paired) and a mix of maternally and paternally derived alleles

Embryology at a Glance, First Edition. Samuel Webster and Rhiannon de Wreede.

Time period: day 0 to adult

Diversity

Cell division by mitosis gives no opportunity for change or diversity, which is ideal for processes like growth and repair. In humans, sexual reproduction allows random mingling of maternal and paternal DNA to produce a new, unique individual. This is able to occur because of a different type of cell division called **meiosis**.

During meiosis a single cell divides twice to form four new cells. These daughter cells have half the normal number of chromosomes (they are **haploid** cells). Meiosis is the method of producing **spermatozoa** and **oocytes**. When an egg is fertilised by a sperm the chromosomes will combine to form a cell with the normal number of chromosomes.

Human chromosomes

There are **23 pairs** of human chromosomes (Figure 6.1) in a normal, **diploid** cell (from the Greek word *diploos*, meaning 'double'). Each chromosome is a length of DNA wrapped into an organised structure (Figure 6.2). Twenty-two of the pairs of chromosomes are known as **autosomes**. The remaining pair are known as the **sex chromosomes**, which hold genes linked to the individual's sex. When condensed the pairs of autosomes look like X's (Figures 6.3 and 6.4), and the sex chromosomes look like X's or Y's (Figure 6.1). The female sex chromosome pair appears as XX, the male as XY.

Meiosis I

A cell dividing by meiosis divides twice (meiosis I and meiosis II). During meiosis I (Figure 6.5), a cell passes through phases very similar to those of mitosis, but with some significant differences. It begins with 23 pairs of chromosomes (46 chromosomes in total).
- **Interphase:** the cell goes about its normal, daily business (**diploid**).
- **Prophase I: homologous chromosomes** exchange DNA (homologous recombination); chromosomes condense and become visible; centrioles move to opposite ends of the cell and extend microtubules out (mitotic spindle); centromeres extend fibres out from chromosomes (**diploid**).
- **Prometaphase I:** the nuclear membrane disappears, microtubules attach centrioles to centromeres and start pulling the chromosomes (**diploid**).
- **Metaphase I:** chromosomes are aligned in the middle of the cell (**diploid**).
- **Anaphase I:** homologous chromosome pairs split, one of each pair (each pair has two chromatids) moving to either end of the cell (**diploid**).
- **Telophase I:** homologous chromosomes reach each end of the cell; new membranes form around the new nuclei for the daughter cells (**diploid**).
- **Cytokinesis:** an actin ring around the centre of the cell shrinks and splits the cell in two (**haploid**).

After meiosis I each cell has 23 chromosomes, and each chromosome has two chromatids. It is therefore haploid.

Homologous recombination

The key event during meiosis I is the separation of homologous chromosomes, rather than the separation of sister chromatids as occurs during mitosis. But what are homologous chromosomes?

Sister chromatids (Figure 6.4) are identical copies of DNA that are attached to one another by the centromere to form the

X-shaped chromosomes that we recognise. So, when sister chromatids are separated into two new cells by mitosis the new cells will be genetically identical.

Homologous chromosomes (Figure 6.4) are the two chromosomes that make up the 'pair' of chromosomes that we talk about in diploid cells. We say that human diploid cells contain 23 pairs of chromosomes. They are homologous in that they are the same chromosome but with subtle differences. One chromosome has been inherited from the father and one from the mother.

Homologous chromosomes contain genes for the same biological features, but the genes may be slightly different. For example, the genes for eye colour would be found on both homologous chromosomes but one chromosome may hold the gene that encodes for blue eyes and the other for green eyes. These are different **alleles** of the same gene.

During homologous recombination those genes are swapped around randomly between the homologous chromosomes before they are pulled into new cells. Therefore, each new cell could be quite different with many, many genes randomly exchanged. In this way the gametes (eggs, sperm) formed by meiosis become very diverse.

The female sex chromosomes (XX) are homologous, but the male sex chromosomes (XY) are not.

Meiosis II

Without replicating its DNA the cell moves from meiosis I to meiosis II. Meiosis II is very similar to mitosis.
- **Prophase II:** chromatids condense and become visible; centrioles move to opposite ends of the cell and extend microtubules out (mitotic spindle); centromeres extend fibres out from chromosomes (**haploid**).
- **Prometaphase II:** the nuclear membrane disappears, microtubules attach centrioles to centromeres and start pulling the chromosomes (**haploid**).
- **Metaphase II:** chromosomes are aligned in the middle of the cell (**haploid**).
- **Anaphase II:** chromosome pairs split (centromeres cut), one of each pair (sister chromatids) moving to either end of the cell (**haploid**).
- **Telophase II: sister chromatids** reach opposite ends of the cell; new membranes form around the new nuclei for the daughter cells (**haploid**).
- **Cytokinesis:** an actin ring around the centre of the cell shrinks and splits the cell in two (**haploid**).

The end result is, generally speaking, 4 cells with 23 unpaired chromosomes each (Figure 6.6). We will find out more about this in the gamete chapters (see Chapter 7, spermatogenesis and Chapter 8, oogenesis).

Clinical relevance

Karyotyping and comparing a patient's chromosomes to the expected normal chromosomal pattern is important in diagnosing a number of chromosomal abnormalities, such as trisomy 21 (Down syndrome), XXY (Klinefelter syndrome) and trisomy 18 (Edwards syndrome).

The homologous recombination of prophase I is an important mechanism of **Mendelian inheritance**. It is a key tenet of modern genetics and underlies most clinical disorders with a genetic basis.

7 Spermatogenesis

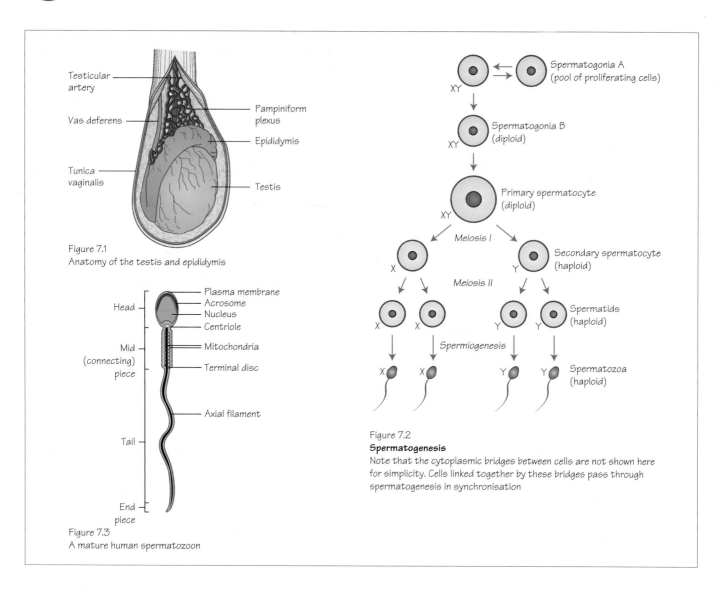

Figure 7.1
Anatomy of the testis and epididymis

- Testicular artery
- Vas deferens
- Tunica vaginalis
- Pampiniform plexus
- Epididymis
- Testis

Figure 7.3
A mature human spermatozoon

- Head
- Mid (connecting) piece
- Tail
- End piece

- Plasma membrane
- Acrosome
- Nucleus
- Centriole
- Mitochondria
- Terminal disc
- Axial filament

Spermatogonia A (pool of proliferating cells) — XY
Spermatogonia B (diploid) — XY
Primary spermatocyte (diploid) — XY

Meiosis I

Secondary spermatocyte (haploid) — X, Y

Meiosis II

Spermatids (haploid) — X, X, Y, Y

Spermiogenesis

Spermatozoa (haploid) — X, X, Y, Y

Figure 7.2
Spermatogenesis
Note that the cytoplasmic bridges between cells are not shown here for simplicity. Cells linked together by these bridges pass through spermatogenesis in synchronisation

Embryology at a Glance, First Edition. Samuel Webster and Rhiannon de Wreede.

Time period: puberty to death

Meiosis continued

In the last chapter we talked about the importance of meiosis in sexual reproduction and diversity, and saw how haploid cells are formed. In males, meiosis occurs during spermatogenesis, in which **spermatogonia** in the testes become **spermatozoa**.

The germ cells that will form the male gametes (spermatozoa) are derived from germ cells that migrate from the yolk sac into the site of early gonad formation (see Chapter 36).

Aims of spermatogenesis

Spermatogonia are diploid germ cells in the testes that maintain their numbers by mitosis, thus maintaining spermatozoa numbers through life. Spermatogonia contain both X and Y sex chromosomes. At a certain point a spermatogonium will stop its other duties and begin meiosis. The cells that result will then pass through more stages of maturation and development and will become mature spermatozoa capable of travelling to and fertilising an ovum.

Anatomy

The testis is made up of very long, tightly coiled tubes called the **seminiferous tubules** that are surrounded by layers of connective tissue, blood vessels and nerves (Figure 7.1). The seminiferous tubules are linked to **straight tubules** and a network of tubes called the **rete testis** which lead to the **epididymis**. The epididymis is another collection of tubes on the posterior edge of the testis that passes inferiorly and is continuous with the **ductus deferens** (also known as the vas deferens). The ductus deferens carries mature spermatozoa from the testis to the urethra.

Spermatogonia are found in the walls of the seminiferous tubules, and as they progress through spermatogenesis they pass towards the lumina of those tubules. Leydig cells within the testes produce testosterone. Sertoli cells are also found in the seminiferous tubules, and produce a number of hormones.

Spermatocytogenesis

The spermatogonia that we begin the process with are called **spermatogonia A** cells (Figure 7.2). These are the stem cells that proliferate and replenish the root source of all spermatozoa. The cells that are about to begin meiosis are called **spermatogonia B** cells, and can be recognised partly because they are connected to one another by cytoplasmic bridges. They continue to divide by mitosis until they become **primary spermatocytes**. The cytoplasmic bridges will maintain connections between a group of cells during spermatogenesis, synchronising the process and batch producing groups of spermatozoa.

The primary spermatocytes enter **meiosis I**. Homologous recombination of chromosomes occurs in this stage. One primary spermatocyte becomes two **secondary spermatocytes**. These cells are now haploid. Each secondary spermatocyte may contain an X or a Y sex chromosome.

Secondary spermatocytes enter **meiosis II** and again divide, forming **spermatids**. As the DNA was not replicated in meiosis II these cells have half their original DNA. During fertilisation this DNA will be combined with the DNA of the maternal ovum. This is the end of the first stage of spermatogenesis, known as **spermatocytogenesis**.

Spermiogenesis

During spermiogenesis the rounded spermatid cell changes shape, becoming elongated and developing the familiar head and tail. The cell loses cytoplasm, the nucleus is packed into the head, mitochondria become concentrated in the first part of the tail and an acrosome forms around the tip of the head. The acrosome contains enzymes that will help the sperm penetrate the outer layers of the ovum during fertilisation.

At the end of spermiogenesis the spermatids have become spermatozoa (Figure 7.3).

Spermatozoa

Spermatogenesis takes around 64 days to produce spermatozoa from germ cells in the above processes. The spermatozoa are then passed in an inactive state to the epididymis, where they continue to mature. During the next week they descend within the epididymis and become motile and ready to be passed into the ductus deferens during ejaculation.

Clinical relevance

Abnormalities in spermatogenesis are common, and during fertility investigations the number and concentration of spermatozoa, and the proportion of abnormal sperm, are counted in a semen sample. A number of biological and environmental factors will affect the sperm count and fertility, such as smoking, sexually transmitted diseases, toxins, testicular overheating and radiation.

8 Oogenesis

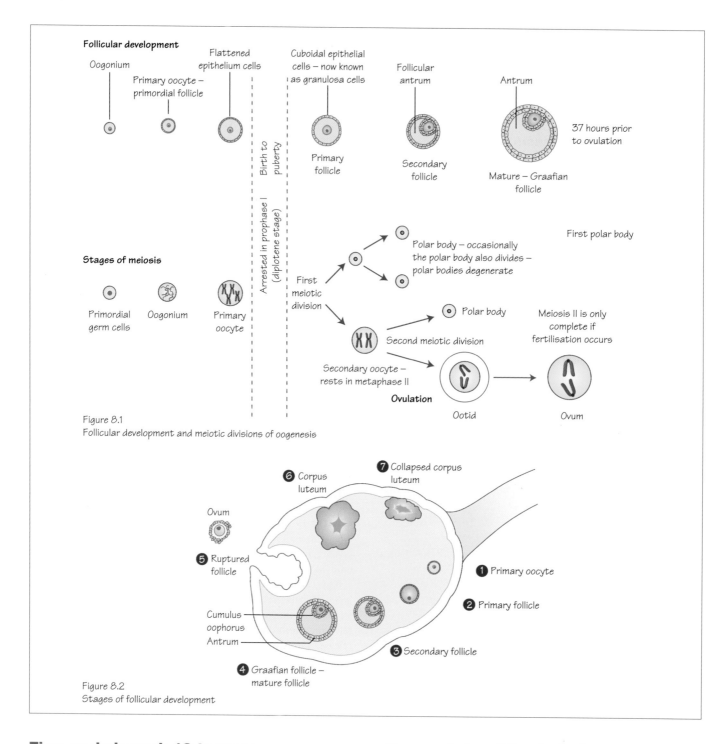

Follicular development

Oogonium

Primary oocyte – primordial follicle

Flattened epithelium cells

Cuboidal epithelial cells – now known as granulosa cells

Primary follicle

Follicular antrum

Secondary follicle

Antrum

Mature – Graafian follicle

37 hours prior to ovulation

Birth to puberty

Arrested in prophase I (diplotene stage)

Stages of meiosis

Primordial germ cells

Oogonium

Primary oocyte

First meiotic division

Polar body – occasionally the polar body also divides – polar bodies degenerate

First polar body

Second meiotic division

Polar body

Meiosis II is only complete if fertilisation occurs

Secondary oocyte – rests in metaphase II

Ovulation

Ootid

Ovum

Figure 8.1
Follicular development and meiotic divisions of oogenesis

❼ Collapsed corpus luteum

❻ Corpus luteum

Ovum

❺ Ruptured follicle

Cumulus oophorus

Antrum

❹ Graafian follicle – mature follicle

❶ Primary oocyte

❷ Primary follicle

❸ Secondary follicle

Figure 8.2
Stages of follicular development

Time period: week 12 to menopause

Overview

Female germ cells proliferate by mitosis in the ovaries to form a large number of **oogonia**. These cells are diploid, contain two X sex chromosomes, and will become haploid **mature oocytes** via the process of oogenesis. This process is similar to spermatogenesis but has some significant differences.

The germ cells that will form the female gametes (oocytes) are derived from germ cells that migrate from the yolk sac into the site of early gonad formation (see Chapter 36).

Ovaries

The ovaries are a pair of organs that produce oocytes and reproductive hormones. They lie near the openings of the **uterine tubes** (also known as the Fallopian tubes or oviducts) that extend from

Embryology at a Glance, First Edition. Samuel Webster and Rhiannon de Wreede.

the uterus. Finger-like projections from the uterine tubes called fimbriae collect oocytes when they are expelled from the ovaries. The oocyte is carried into and along the uterine tube for fertilisation and subsequent implantation into the wall of the uterus.

The adult ovary is predominantly made up of connective tissue that supports a large number of follicles. Blood vessels and nerves are concentrated within the central medulla whereas follicles are found in the outer cortex, in varying stages of development.

Meiosis I

Oogonia begin oogenesis by entering meiosis I in week 12 of embryonic development (Figure 8.1). During meiosis I the cell is known as the **primary oocyte**, and is surrounded by a thin layer of squamous epithelial cells. This structure is a **follicle**, and in its very early stage is called a **primordial follicle**. The primary oocyte at this stage is developmentally arrested in prophase of meiosis I. This pause in development may continue for 45 years or more.

The number of primordial follicles vastly increases during the foetal period but many degenerate, leaving around 400,000 follicles available at puberty. After birth no new oocytes form.

Puberty

With the onset of puberty some of the stalled primary oocytes continue oogenesis each month (Figure 8.1).

The primary oocyte becomes larger and the follicular cells around it become cuboidal and the layer thickens. The follicle is now a **primary follicle** (Figure 8.2).

The oocyte and the granulosa (follicle) cells produce a layer of glycoproteins on the surface of the oocyte called the **zona pellucida**.

When the follicle forms more than one layer of granulosa cells it is called a **secondary follicle**.

One follicle continues to develop and grow, and the others degenerate. It is not clear how one follicle is chosen over the others.

A cavity called the **antrum** forms between the layers of granulosa cells, and the mass of follicular cells is now termed the **cumulus oophorus**.

The connective tissue cells of the ovary around the follicle respond by differentiating and forming two new layers: the **theca interna** and the **theca externa**. The theca interna has a hormonal role, and the theca externa a supportive role. This follicle is now a **mature vesicular follicle** or **Graafian follicle**.

The thecal and granulosa cells of the developing follicles produce oestrogens that cause the thickening of the endometrial lining of the uterus and other preparations for receiving a fertilised oocyte. This occurs from days 5 to 14 of the menstrual cycle (see Figure 4.1 and Chapter 11).

Ovulation

The primary oocyte of the Graafian follicle responds to surges in **follicle stimulating hormone** (FSH) and **luteinising hormone** (LH) produced by the pituitary gland on days 13–14 of the menstrual cycle by resuming meiosis I and continuing its stalled cell division (Figure 8.1).

When the oocyte divides it forms one large cell and one smaller remnant of the division known as a **polar body**. At the end of meiosis I the oocyte has become a **secondary oocyte**.

Polar bodies

Polar bodies are small, non-functional cells. They receive very little of the available cytoplasm and degenerate soon after division. In this way the oocyte is able to retain its size but discard chromosomal material to become a haploid cell ready for fertilisation.

One polar body is formed with meiosis I and two polar bodies are formed with meiosis II.

Meiosis II

The secondary oocyte begins meiosis II but this division is again halted, this time during metaphase II. Meiosis II will only continue if the oocyte is fertilised.

Post-ovulation

With ovulation the secondary oocyte is passed into the uterine tube, but the follicle remains within the ovary (Figure 8.2). At this stage the follicle is very large and makes up a significant portion of the ovary. This follicle becomes the **corpus luteum**.

In response to LH the corpus luteum produces progesterone, oestrogens and other hormones causing the endometrium of the uterus to thicken further, develop its vasculature, form glands and prepare for implantation.

If fertilisation does not occur the corpus luteum degenerates about 14 days later and becomes a scar tissue remnant of itself called the **corpus albicans**. Hormone production ceases and menstruation begins as the thickened endometrium is shed.

Clinical relevance

The primary oocyte may be arrested in meiosis I throughout life for 40–50 years if it is not triggered to continue development until a menstrual cycle late in reproductive life. DNA fragmentation within those stored oocytes is more common in older women as DNA damage increases with time. This may be the reason for reduced fertility with increasing age.

Knowledge of the sex hormones' effects on follicle development have allowed the invention of the oral contraceptive pill. High levels of oestrogens and progesterone inhibit gonadotrophin releasing hormone (GnRH) and subsequently LH and FSH release. Decreased levels of FSH mean that the follicle is not stimulated to develop, and the absence of an LH surge prevents ovulation occurring.

Chemotherapy and radiotherapy can destroy primordial ovarian follicles. As there is a finite reserve of oocytes formed prenatally, which cannot be replenished after treatment, the cryopreservation of oocytes before treatment begins should be considered. Frozen oocytes may be used for *in vitro* fertilisation at a later date if the patient becomes infertile.

9 Fertilisation

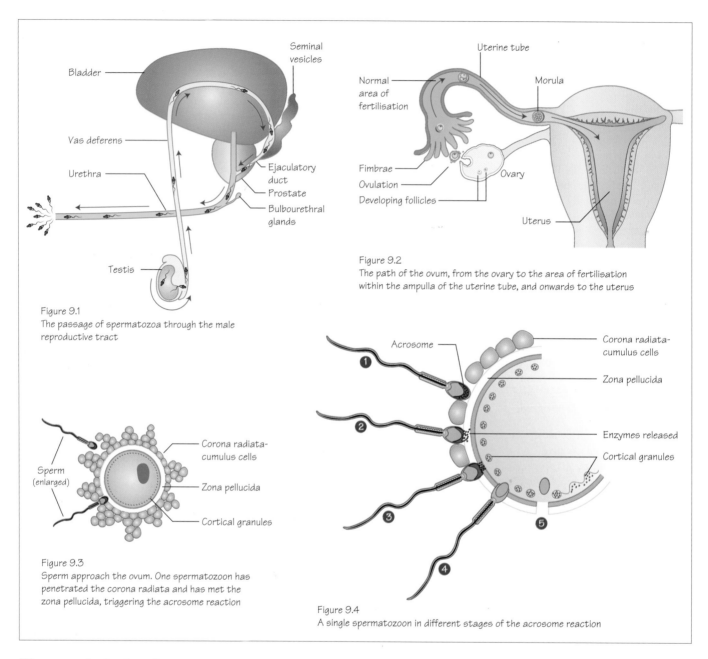

Figure 9.1
The passage of spermatozoa through the male reproductive tract

Figure 9.2
The path of the ovum, from the ovary to the area of fertilisation within the ampulla of the uterine tube, and onwards to the uterus

Figure 9.3
Sperm approach the ovum. One spermatozoon has penetrated the corona radiata and has met the zona pellucida, triggering the acrosome reaction

Figure 9.4
A single spermatozoon in different stages of the acrosome reaction

Time period: day 0

Fertilisation

With meiosis and sexual reproduction an organism is able to reproduce and create genetically individual offspring. Here we discuss what happens when the gametes (ovum and spermatozoon) meet, combine their genetic material and begin the formation of an embryo.

Capacitation

Spermatozoa in the female genital tract become prepared for fertilisation with a process called **capacitation**. Spermatozoa stored in the epididymis pass through the ductus deferens during ejacula-

tion and mix with secretions from the seminal vesicles, prostate and bulbourethral glands (Figure 9.1) as they are released into the vagina. With this, and a possible cue from the female environment, the outer surface of the acrosome becomes modified by the removal of glycoproteins and proteins. This is the final maturation step of the spermatozoa.

The spermatozoa become hyperactive and make their way through the cervix, uterus and uterine tube to find the ovum.

Ovulation

With ovulation the secondary oocyte (or ovum) is expelled from the follicle on the ovary surface. Fimbriae at the opening of the uterine tube collect it and pass it into the uterine tube (Figure 9.2).

Embryology at a Glance, First Edition. Samuel Webster and Rhiannon de Wreede.

The ovum is moved towards the ampulla of the uterine tube where it has roughly 24 hours to meet with a spermatozoon to become fertilised.

Acrosome reaction

The oocyte is surrounded by **cumulus** cells (also termed the corona radiata) from the follicle and spermatozoa must break through this outer layer to reach the oocyte itself (Figure 9.3). When a spermatozoon succeeds in this it encounters the **zona pellucida** surrounding the plasma membrane of the oocyte and insulating it from the external environment. The spermatozoon binds to the zona pellucida and is triggered to begin the acrosome reaction.

The acrosomal cap of the head of the sperm breaks down, releasing enzymes that dissolve the zona pellucida locally allowing the spermatozoon to enter the oocyte (Figure 9.4).

Cortical reaction

Once through the zona pellucida the membranes of the egg and sperm meet and fuse. The contents of the sperm are now within the egg, as its plasma membrane is left behind and lost (Figure 9.4).

Cortical granules containing enzymes are released from the egg, causing the binding proteins of the entire zona pellucida to become altered, preventing further sperm from binding.

With the zona pellucida, and the acrosome and cortical reactions fertilisation by multiple sperm (dispermy or polyspermy) is prevented. This is a very important process in mammalian reproduction as hundreds of sperm reach the egg at the same time and dispermy would create an embryo with three haploid sets of chromosomes (triploidy) that would be extremely unlikely to survive.

Meiosis II

The secondary oocyte was paused partway through meiosis II (see Chapter 8). With the fusion of the spermatozoon cell membrane the oocyte is triggered to continue meiosis.

The two cells that result from this division are the **definitive oocyte** and the second polar body. The second polar body receives little cytoplasm, allowing the definitive oocyte to maintain its size.

Zygote

The fertilised oocyte contains the DNA of the spermatozoon and the DNA of the oocyte. In principle it contains a diploid set of chromosomes.

Although the DNA has not been reorganised yet, fertilisation has formed a genetically unique individual. This cell can be called a **zygote** (see Chapter 10).

Mitosis and DNA

The spermatozoon's nucleus becomes the male **pronucleus**, and aligns with the female pronucleus. Each pronucleus is haploid at this stage. The two pronuclei lose their nuclear membranes and their DNA is duplicated. This takes around 18 hours.

The DNA condenses into chromosomes, and paternal and maternal chromosomes become aligned together on the equator of the cell. Sister chromatids from each chromosome are pulled towards either end of the cell, as observed during anaphase in the mitosis chapter (see Figure 5.4).

Mitosis continues and the cell is split in two.

Chromosomes

With fertilisation the diploid number of chromosomes has been restored by combining chromosomes from the father and the mother.

The spermatozoon will bring either an X or Y sex chromosome to the oocyte's X sex chromosome. The spermatozoon determines the sex of the embryo by producing either an XY (male) or XX (female) pair of sex chromosomes.

Embryological and clinical timings

Fertilisation occurs during an 18–24 hour period shortly after ovulation in humans. It is impossible to determine an exact time of fertilisation, and very difficult to determine on which day fertilisation occurred.

Embryologically we talk about developmental processes occurring a number of days after fertilisation. For example, we say that the first somites form at 20 days. These are the timings that we use in this book, and that appear at the top of each chapter.

Clinically, however, gestation is timed from a more evident event: the last menstrual period (LMP). As ovulation occurs fairly reliably 2 weeks after menstruation, and fertilisation occurs within 24 hours of ovulation, it is easier and more reliable to note the date of the LMP for a patient and from this record weeks of pregnancy and the predicted date of birth.

It is important to be aware that there is a 2-week difference between embryological and clinical timings (see Figure 4.2). If this textbook notes that the first somites occur at 20 days (around 3 weeks after fertilisation), this occurs at 5 weeks clinically.

Clinical relevance

An **extra-uterine** pregnancy (or ectopic pregnancy) can occur because of the movement of the ovum from the ovaries to the uterus. A fertilised ovum may implant into the uterine tube, the cervix, the ovary or the abdomen. Tubal pregnancies within the uterine tube are the most common type. Typically, an ectopic pregnancy is not viable and in extreme cases can lead to the death of the mother.

For *in vitro* **fertilisation** techniques, sperm must be artificially induced to begin capacitation. With capacitation the sperm is primed to undergo the acrosome reaction when it meets the ovum.

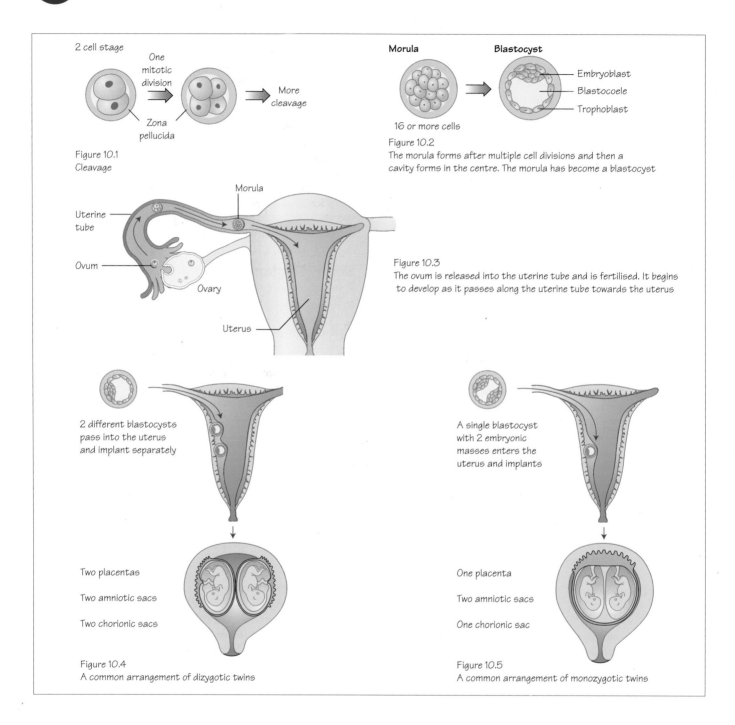

2 cell stage

One mitotic division

Zona pellucida

More cleavage

Figure 10.1
Cleavage

Morula   **Blastocyst**

Embryoblast
Blastocoele
Trophoblast

16 or more cells

Figure 10.2
The morula forms after multiple cell divisions and then a cavity forms in the centre. The morula has become a blastocyst

Morula

Uterine tube

Ovum

Ovary

Uterus

Figure 10.3
The ovum is released into the uterine tube and is fertilised. It begins to develop as it passes along the uterine tube towards the uterus

2 different blastocysts pass into the uterus and implant separately

Two placentas

Two amniotic sacs

Two chorionic sacs

Figure 10.4
A common arrangement of dizygotic twins

A single blastocyst with 2 embryonic masses enters the uterus and implants

One placenta

Two amniotic sacs

One chorionic sac

Figure 10.5
A common arrangement of monozygotic twins

Time period: days 0–5

Zygote

With fertilisation the oocyte and spermatozoon combine to become a zygote. The zygote is the simplest form of the new animal, and will begin to split and divide into new cells that will become organised, specialised and form shapes and new structures as it becomes more complex.

Cleavage

Around 24 hours after fertilisation the zygote begins to increase its number of cells by rapid mitosis, but without increasing its size. The cells become smaller with each cell division. The number of cells doubles with each division. This is **cleavage** (Figure 10.1).

The cells of the zygote are called **blastomeres**.

Morula

Cells become compacted and tightly squashed together. From around the 12-cell stage the ball of cells becomes called the **morula** (Figure 10.2), derived from the Latin word for mulberry, which it now resembles.

The cells of the morula will not only give rise to the cells of the embryo, but also to many of its supporting structures, such as part of the placenta.

By this stage the cells are communicating with each other and becoming organised and ready for the next stage.

The blastomeres in the middle of the morula become the **inner cell mass** or **embryoblast**. These cells will directly form the embryo.

The blastomeres on the outside of the morula become the **outer cell mass** or **trophoblast**. These cells will form some of the supporting structures for the embryo.

Blastocyst

The morula passes into the uterus around 4 days after fertilisation (Figure 10.3).

Trophoblast cells pull luminal fluid from the uterine cavity into the centre of the morula (Figure 10.2). The fluid-filled space that forms is called the **blastocoel** (or blastocyst cavity). The cells of the inner cell mass are pushed to one end of the cavity and become called the **embryonic pole**. The morula is now called a **blastocyst**.

Implantation

Around 5 days after fertilisation the blastocyst loses the **zona pellucida**. By doing this it becomes able to grow in size and interact with the uterine wall. The blastocyst attaches to the endometrial epithelium lining the uterus, triggering changes to the trophoblast and to the endometrium in preparation for the implantation of the blastocyst into the uterine wall (see Chapter 12).

Twins

Twinning can occur in different ways. Two separate blastocysts may form from fertilisation by different sperm of two different ova released from an ovary simultaneously. These twins would not be identical twins, and they would have separate placentas (dichorionic), separate amniotic sacs (diamniotic) and may even be of different sexes (Figure 10.4). These would be **dizygotic twins** (or fraternal or non-identical twins).

A zygote may split during cleavage, or later, when the inner cell mass has formed, or later still, when the embryo has become more complicated and formed a bilaminar embryonic disc (see Chapter 12). If the zygote splits during cleavage each blastocyst will implant separately. If the zygote splits at a later stage the two embryos may share the same chorion, amnion or placenta (Figure 10.5).

If a single zygote splits identical twins will grow. These twins would come from the same ovum and spermatozoon, so would be genetically identical. These would be **monozygotic twins** (or identical twins). This is rarer. It is common for monozygotic twins to share a placenta (monochorionic), but have separate amniotic sacs (diamniotic). This situation arises from cleavage of the blastocyst 4–8 days after fertilisation. A small number of monozygotic twins share their amnion (monoamniotic), and this occurs if the division of the zygote occurs later than 9 days after fertilisation. The more tissues shared between twins the greater the risk to the embryos. Hence, dizygotic twins have the lowest mortality risk. Conjoined twins are at significant risk. This situation arises when the zygote splits incompletely later than 12 days after fertilisation.

Clinical relevance

It is thought that blastocyst abnormalities are common and not compatible with life. Most probably do not implant into the uterus, show no signs of pregnancy and therefore often the pregnancy is not detected.

Twins are more likely to be born prematurely, resulting in low birth weights and the associated complications.

In vitro fertilisation (IVF) treatments can result in multiple zygotes because of the drugs used to encourage ovulation. Clomid is a drug that blocks oestrogen receptors, so the body perceives low oestrogen levels, more FSH is released and more follicles mature in the ovary to be released and fertilised.

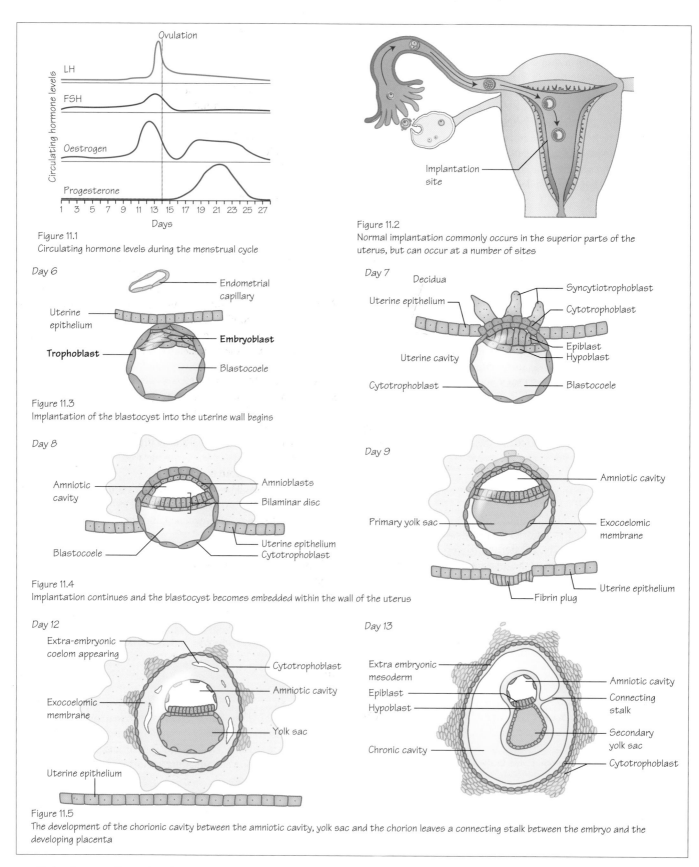

Figure 11.1
Circulating hormone levels during the menstrual cycle

Figure 11.2
Normal implantation commonly occurs in the superior parts of the uterus, but can occur at a number of sites

Figure 11.3
Implantation of the blastocyst into the uterine wall begins

Figure 11.4
Implantation continues and the blastocyst becomes embedded within the wall of the uterus

Figure 11.5
The development of the chorionic cavity between the amniotic cavity, yolk sac and the chorion leaves a connecting stalk between the embryo and the developing placenta

Embryology at a Glance, First Edition. Samuel Webster and Rhiannon de Wreede.

Time period: days 5–13

Introduction
At this early stage of development timings are very individual and often a range is more appropriate. All timings described here are the typical ages documented in a range of literature.

Implantation
The travelling morula enters the uterus at approximately day 4 and begins to form a blastocyst at around 4.5 days. It begins the process of implantation roughly a day later (6–7 days). Implantation occurs to enable the developing embryo to take oxygen and nutrients from the mother, thus enabling its growth.

The menstrual cycle (uterus)
For the blastocyst to implant successfully the walls of the uterus make certain preparations. The endometrial lining of the uterus undergoes changes every month as part of the menstrual cycle. There are three main stages: the **proliferative (or follicular) phase**, the **secretory (luteal or progestational) phase** and the **menstrual phase**.

The **proliferative phase** begins on day 5 of an average menstrual cycle and finishes on day 13, prior to ovulation. Changes to the uterus during this phase include an increase in thickness of the endometrium and an increase in vascularisation.

During the **secretory phase** arteries and glands become coiled and secretions increase helping to maintain the thickness of the endometrium. There are three distinct layers of the endometrium: a superficial **compact** layer, a middle **spongy** layer and a deep **basal** layer.

There is also a group of new rounded cells that cover the whole surface of the endometrium. These are the **decidual cells**. If fertilisation does not occur the spongy and compact layers and the decidual cells are shed.

Decidualization
Decidualization is the collective term for the changes that the endometrium undergoes in pregnancy. Decidual cells have a high secretory capacity of laminin and fibronectin (both have adhesive qualities) and the vascularity of the tissue is improved. At implantation these cells accumulate fats and glycogen.

The decidua remains important and has a role in the development of the placenta (see Chapter 12).

The menstrual cycle (hormones)
The phases of the menstrual cycle are coordinated by hormones (Figure 11.1).

Proliferative (follicular) phase
Follicle stimulating hormone (FSH), secreted from the anterior lobe of the pituitary gland, initiates the maturation of a few follicles in the ovary. A day before ovulation the pituitary gland also releases a surge of **luteinising hormone** (LH) inducing the ovary to release an ovum.

The empty **Graafian follicle** (see Chapter 8) matures and produces oestrogen. In a positive feedback loop the oestrogen induces more FSH and LH to be produced by the pituitary and consequently more oestrogen is produced. This causes the thickening of the endometrium.

Secretory (luteal) phase
The Graafian follicle matures into the **corpus luteum** and begins to produce progesterone as well as oestrogen. Progesterone maintains the developing endometrium and increases the uterine gland secretions. The presence of progesterone and oestrogen inhibits production of FSH and LH dropping the levels of both hormones.

Upon fertilisation and implantation the trophoblast cells produce **human chorionic gonadotrophin hormone** (hCG) causing the corpus luteum to continue progesterone production.

Implantation mechanism
The location for implantation is commonly superiorly on the anterior or posterior walls of the uterus (Figure 11.2).

At implantation the blastocyst comprises a fluid-filled core, an outer cell mass (trophoblast) and an inner cell mass (embryoblast) at the embryonic pole (Figure 11.3).

The process of implantation can be broken down into four stages. The first is **hatching**, as the developing blastocyst has to 'hatch' out of its surrounding **zona pellucida**. **Apposition** follows, as the trophoblast cells come into contact with the decidua of the endometrium. If the embryonic pole is not closest to the area of contact the inner cell mass rotates to become aligned with the decidua. Then **adhesion** occurs and molecular communication between blastocyst and endometrial cells is vastly increased. Finally, **invasion** of the endometrium by the trophoblast begins.

Bilaminar germ disc
By day 8 implantation has begun and the blastocyst develops again into a more complex structure. The inner cell mass differentiates into an **epiblast** layer and a **hypoblast** layer (Figure 11.4). The hypoblast layer is located nearer to the blastocyst cavity. These two layers are now called the **bilaminar disc**.

Simultaneously another cavity forms within the epiblast, called the **amniotic cavity**. The cells of the hypoblast will develop into the extraembryonic membranes (amnion, yolk sac, chorion and allantois) and the epiblast will develop to form the embryo (Figure 11.5).

Clinical relevance
Implantation can be negatively influenced by many factors at any stage of the process. This is a common cause of miscarriage, especially for couples undergoing IVF treatment.

Immunosuppressant cytokines are produced during implantation to prevent an immune reaction, and some **autoimmune diseases** (systemic lupus erythematosus and antiphospholipid syndrome) can mean the mother's body attacks the embryo at implantation.

If implantation occurs near to the internal os of the cervix the placenta can develop in a dangerous position (**placenta praevia**) which can result in severe bleeding in later pregnancy and labour.

Intrauterine devices (IUDs) used for contraception were originally intended to prevent implantation of the blastocyst by irritating the endometrium. It is likely that instead they work by inhibiting sperm and ovum migration and fertilisation. Medicated IUDs also contain progesterone, which inhibits FSH and LH release, preventing ovulation.

12 Placenta

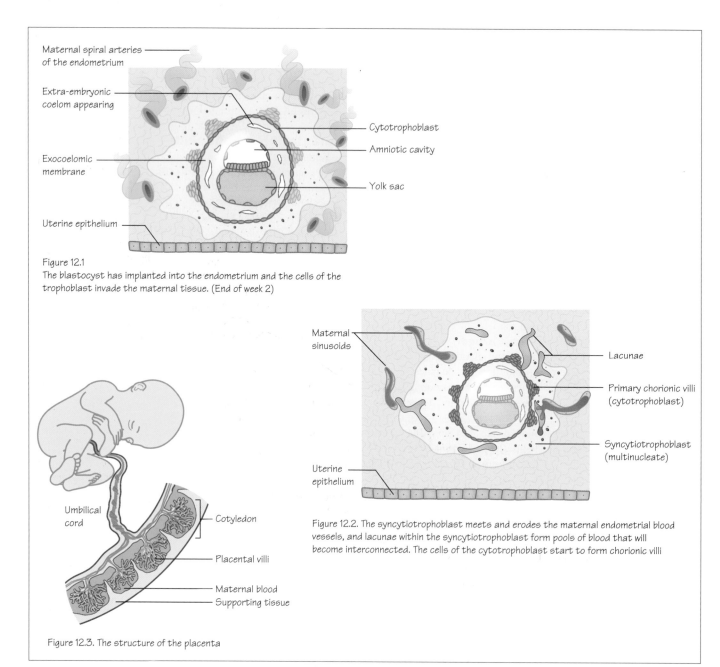

Figure 12.1
The blastocyst has implanted into the endometrium and the cells of the trophoblast invade the maternal tissue. (End of week 2)

Labels (Figure 12.1):
- Maternal spiral arteries of the endometrium
- Extra-embryonic coelom appearing
- Exocoelomic membrane
- Uterine epithelium
- Cytotrophoblast
- Amniotic cavity
- Yolk sac

Figure 12.2. The syncytiotrophoblast meets and erodes the maternal endometrial blood vessels, and lacunae within the syncytiotrophoblast form pools of blood that will become interconnected. The cells of the cytotrophoblast start to form chorionic villi

Labels (Figure 12.2):
- Maternal sinusoids
- Uterine epithelium
- Lacunae
- Primary chorionic villi (cytotrophoblast)
- Syncytiotrophoblast (multinucleate)

Figure 12.3. The structure of the placenta

Labels (Figure 12.3):
- Umbilical cord
- Cotyledon
- Placental villi
- Maternal blood
- Supporting tissue

Time period: day 7 to week 12

Introduction

As the human embryo grows its need for nutrition increases, requiring a connection to the mother for nutrient, gas and waste exchange. The placenta develops to meet these needs.

Trophoblast

The **trophoblast** develops from the outer layer of the blastocyst before implantation into the endometrium. Trophoblast cells produce **human chorionic gonadotrophin** (hCG). Around 6–7 days after fertilisation the trophoblast begins to invade the endometrium, triggering the decidual reaction (see Chapter 11) and the process that will form the placenta from both embryonic and maternal tissues (see Figure 11.3).

The trophoblast layer has important roles in implantation and placental development, and protects the embryo from maternal immunological attack. With implantation the trophoblast divides into two layers (Figure 12.1): the inner **cytotrophoblast** (mononuclear cells) and the outer **syncytiotrophoblast** (multinucleated).

Embryology at a Glance, First Edition. Samuel Webster and Rhiannon de Wreede.

After 2 weeks the front line of invading trophoblasts of the syncytiotrophoblast reach the endometrial blood vessels and erode them, forming pools of maternal blood within trophoblastic lacunae that have formed (Figure 12.2). At the same time **chorionic villi** begin to grow from embryonic tissue, and will grow, branch and become more complex until the end of the second trimester.

Initially, chorionic villi cover the whole surface of the chorion and by the end of the third week embryonic blood begins to flow through the capillaries within the villi. A week after the chorionic villi appear the basic structure of the placenta has formed and the embryo has developed a primitive circulatory system (see Chapter 25).

Structure

Development of the placenta continues to give a mature placental structure at 12–13 weeks.

Villi become restricted to just one region of the chorion. Linked pools of maternal blood are filled by spiral arteries of the endometrium, themselves branches of the uterine arteries. Foetal blood enters the placenta through the two umbilical arteries, which branch and continually divide until they reach the looping capillaries of the chorionic villi. These branching blood vessels form 15–25 units called **cotyledons** (Figure 12.3).

In the villi the syncytiotrophoblasts and endothelium create the barrier between maternal and fetal blood. Due to the syncytiotrophoblasts' multinucleated structure the nuclei gather in certain places (proliferation knots) leaving other areas free of nuclei. These are **exchange zones** and they create an extremely thin and efficient selective barrier through which nutrients, gas, waste products and antibodies may pass.

The villi are bathed in maternal blood, and exchange takes place. Blood returns to the mother through the uterine veins, and the blood within these maternal pools is replaced 2–3 times per minute. Oxygen-rich blood is returned to the foetus by the umbilical vein (see Chapter 29).

Function

The placenta has vital roles in hormone production, nutrient, metabolite and gas exchange, and in protecting the foetus from immune attack by maternal cells and pathogens, and in enabling the passage of antibodies from mother to foetus.

Gas exchange

Oxygen diffuses into the embryonic circulation and carbon dioxide diffuses out. Foetal haemoglobin (HbF) has a higher affinity for oxygen than adult haemoglobin.

Nutrients

For example, amino acids, lipids, glucose, cholesterol and water-soluble vitamins.

Waste removal

For example, urea, bilirubin and creatine.

Hormones

HCG is produced by the placenta for the first 2 months of pregnancy, maintaining the corpus luteum, which in turn produces progesterone to maintain the endometrium. By week 16 the placenta takes on the task of progesterone production.

The placenta also produces oestrogens that aid development of the uterus and mammary glands, and **human chorionic somatomammotrophin** (hCS, or placental lactogen), an insulin antagonist, that modulates maternal carbohydrate metabolism, prioritises foetal access to maternal blood glucose and aids breast development for lactation.

Antibodies

Maternal immunoglobulins are selectively transferred from about 14 weeks, and the foetus gains passive immunity that persists in the newborn infant for several months. Other maternal proteins are degraded by the trophoblast.

Changes to the placenta

Late in the third trimester the syncytiotrophoblast layer develops grape-like nucleated clusters within its cytoplasm called **syncytial knots** which break off and pass into the maternal circulatory system. Shortly before birth **fibrinoid deposits** appear on the villi.

After birth blood flow ceases through the umbilical arteries and veins, and blood flow to the lungs increases as they fill with air. The lungs are about 15 times better at gas exchange than the placenta. The placenta is extruded as the **afterbirth**.

Clinical relevance

Most drugs (infamously, thalidomide), antibiotics and corticoids, some viruses (e.g. toxoplasma, HIV) and other pathogens can pass across the placenta into foetal blood.

Rhesus (Rh) factors are red blood cell surface molecules that will provoke an immune response (against Rh factors). If an Rh– mother bears an Rh+ child her immune system is likely to only see the Rh factors during birth when foetal blood may cross the placenta to meet maternal blood. The mother will develop anti-Rh antibodies. If she bears a second Rh+ child those anti-Rh antibodies will cross the placenta and destroy foetal red blood cells causing erythroblastosis fetalis.

Pre-eclampsia is often diagnosed by increased blood pressure and proteinuria. It occurs in up to 10% of pregnancies and is more common in first-time mothers. It may arise from a shallowly implanted placenta becoming hypoxic and initiating an immune response from the mother. Birth of the baby is the only treatment option.

Placenta accreta, placenta increta and **placenta percreta** involve the placenta attaching too firmly to the wall of the uterus. Accreta is too firmly attached, increta is even more firmly attached (into the myometrium) and percreta is attached through the uterine wall sometimes to internal organs, even as far as the bladder. Manual exploration and the removal of the retained placental tissue are necessary.

Placental insufficiency and **intrauterine growth restriction** (IUGR) describe conditions in which the placenta cannot supply the necessary nutrients to the foetus. Drug or alcohol abuse, smoking, pre-eclampsia, long-term high blood pressure, infections, diabetes, problems with kidney function or Rh incompatibility are all thought to be related.

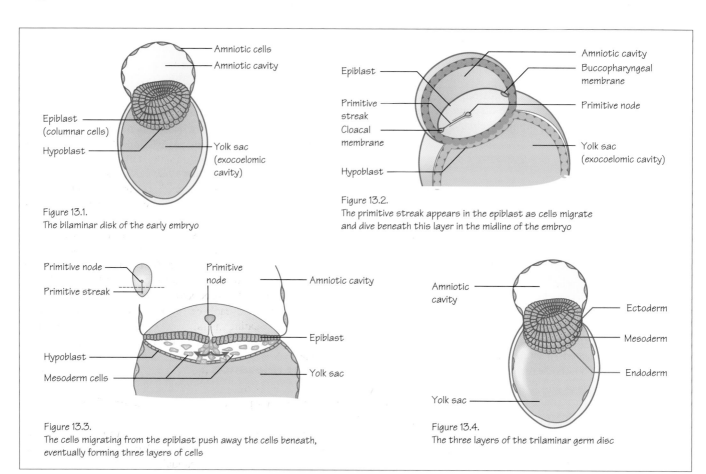

Figure 13.1.
The bilaminar disk of the early embryo

Figure 13.2.
The primitive streak appears in the epiblast as cells migrate
and dive beneath this layer in the midline of the embryo

Figure 13.3.
The cells migrating from the epiblast push away the cells beneath,
eventually forming three layers of cells

Figure 13.4.
The three layers of the trilaminar germ disc

Embryology at a Glance, First Edition. Samuel Webster and Rhiannon de Wreede.

Time period: day 14

Trilaminar disc

If we consider the second week of development produces the **bilaminar disc** (Figure 13.1), we might say that the main event of the third week of development is the formation of the **trilaminar disc**. The process by which this takes place is called **gastrulation**.

The purpose of gastrulation is to produce the three germ layers from which embryonic structures will develop: **ectoderm, mesoderm** and **endoderm**.

Primitive streak

Gastrulation is initiated at about day 14 or 15 with the formation of the **primitive streak** (Figure 13.2). The primitive streak runs as a depression on the epiblastic surface of the bilaminar disc and is restricted to the caudal half of the embryo. Towards the cephalic end there is a round mound of cells called the **primitive node**, surrounding the **primitive pit**.

The appearance of the primitive streak gives the observer an indication of the body axes that the cells are using to organise themselves. Until this point it was unclear which parts of the embryonic sheets were cephalic or caudal (superior or inferior in the adult), ventral or dorsal (anterior or posterior) and left or right. With the primitive streak the embryologist can determine where the head and tail will develop, which side is the left side and which surface will form the outermost layers of the skin.

Epiblast cells migrate towards the streak and when they reach it they invaginate or slip under the epiblast layer to form new layers (Figure 13.3). The first cells to invaginate replace the hypoblast layer and produce the **endodermal layer**.

Some epiblast cells form the **mesodermal layer** between the epiblast layer and the endodermal layer. Cells migrating through the lateral part of the primitive node and cranial part of the streak become **paraxial mesoderm**, cells migrating through the mid-streak level become **intermediate mesoderm** and cells that migrate through the caudal part of the streak are destined to be **lateral plate mesoderm** (see Chapter 23). Cells that migrate through the most caudal tip of the streak contribute to the extra-embryonic mesoderm, along with the cells of the hypoblast.

The epiblast layer now becomes the **ectodermal layer** (Figure 13.4).

After cells have migrated through the streak and begun their path to specialisation, they continue to travel to different areas of the embryo. The first cells that travel towards the cephalic end form the **prechordal plate**, inferior to the **buccopharyngeal** (or oropharyngeal) **membrane**.

The buccopharyngeal membrane will eventually become the mouth opening. Here there is no mesodermal layer; the ectoderm and endoderm are in direct contact. This also occurs at the **cloacal membrane**, which will become the opening of the anus.

Signalling

This period of development is a good example of how the cells of the developing embryo are organised (see Chapter 3). Signalling molecules are a key part of this organisation. There are three groups of molecules involved in the control of our developing embryo: **transcription factors, signalling molecules** and **cell adhesion molecules** (CAMs).

Transcription factors act upon the cells that produce them and affect gene expression by binding DNA and controlling transcription of DNA to mRNA.

A signalling molecule secreted by a cell can affect other cells nearby or at a distance, or the cell that produces it. A cell must have an appropriate receptor ligand to be able to respond to a signalling molecule, and the affect may be positive (e.g. proliferation) or negative (e.g. apoptosis). Signalling molecules are inducers of a wide range of cellular events. Growth factors are a well-known group of signalling molecules.

CAMs allow cells to recognise similar cells or extracellular matrix structures, and aggregate. There are two main groups: calcium dependent (e.g. cadherins) and calcium independent (e.g. integrins).

Often these three types of signalling work in combination to create the complex structures we see develop in morphogenesis. Cells of the primitive streak produce **fibroblast growth factor 8** (a signalling molecule) and this molecule causes a down-regulation in **E-cadherin** (a CAM) production that usually make the cells sticky. Having less E-cadherin means that the cells are more motile, thus stimulating migration towards the primitive streak.

Transcription factors **brachyury** (which acts more dorsally) and **goosecoid** (which activates **chordin**, a signalling molecule) are known to be involved in the differentiation of migrating cells from epiblast to mesoderm.

Also **nodal**, a signalling molecule of the transforming growth factor β (TGF-β) family, is a mesoderm inducer and helps to maintain the primitive streak. An antagonist to nodal called **cerberus** is produced by cells of the hypoblast and thought to cause restriction of the streak at the caudal end of the embryo.

A range of factors are now in play, and the organisation of the embryo is becoming more complicated as it takes shape.

Clinical relevance

Gastrulation is a period of development very susceptible to **teratogens**. In week 3 of development (often before the mother knows of the pregnancy), factors that can have damaging effects on the embryo include **alcohol, caffeine** and **tobacco**. Other known factors that may affect cells at this stage include drugs such as thalidomide, temazepam, forms of retinoic acid (vitamin A), radiation, infections (e.g. rubella and herpes virus) and metabolic imbalances including folic acid deficiency and diabetes. If the embryo is exposed to these factors the upset to signalling or proliferation at an early stage in development results in defects that can be wide ranging and affect multiple developmental processes. Often, the defect originates from a lack of cell numbers in a certain region, and may be so catastrophic as to cause spontaneous abortion.

Sacrococcygeal teratomas occur when cells of the primitive streak get left behind in the sacrococcygeal region, and these cells develop into tumours. Often identified before birth with routine ultrasound scans, most are external and can be removed surgically.

14 Germ layers

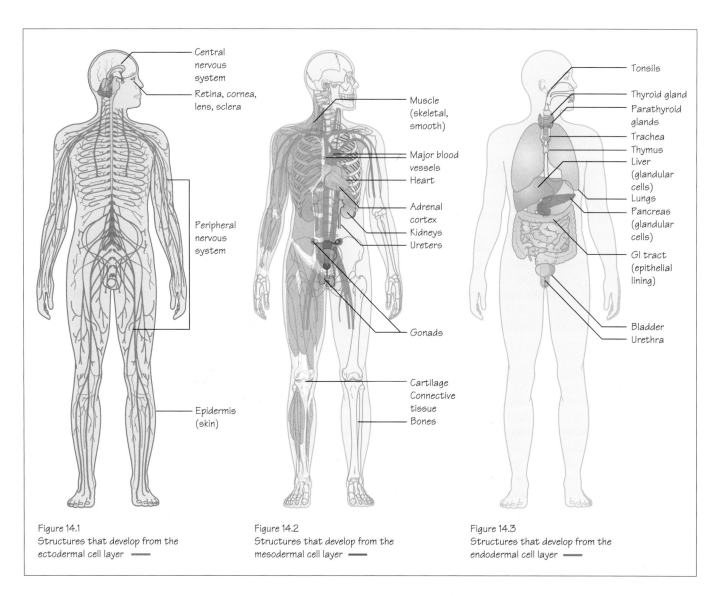

Figure 14.1
Structures that develop from the ectodermal cell layer ——

Central nervous system
Retina, cornea, lens, sclera
Peripheral nervous system
Epidermis (skin)

Figure 14.2
Structures that develop from the mesodermal cell layer ——

Muscle (skeletal, smooth)
Major blood vessels
Heart
Adrenal cortex
Kidneys
Ureters
Gonads
Cartilage
Connective tissue
Bones

Figure 14.3
Structures that develop from the endodermal cell layer ——

Tonsils
Thyroid gland
Parathyroid glands
Trachea
Thymus
Liver (glandular cells)
Lungs
Pancreas (glandular cells)
GI tract (epithelial lining)
Bladder
Urethra

Embryology at a Glance, First Edition. Samuel Webster and Rhiannon de Wreede.

Time period: day 15

Trilaminar disc

In the third week of development the embryonic trilaminar disc is formed, giving the embryo three germ layers: **ectoderm**, **mesoderm** and **endoderm** (see Figure 13.4). From these germ layers almost all of the structures of the embryo will develop.

Ectoderm

The ectoderm will form the external surface of the embryo: the epidermis of the skin (Figure 14.1). The dermis is formed from the mesoderm layer. Melanocytes, the cells that give the skin its pigment, are derived from neural crest cells. These cells are themselves ectodermal and are involved in the development of a range of structures (see Chapters 15 and 16).

The nervous system is also formed from ectoderm (Figure 14.1), as we see when we study neurulation (see Chapter 15). This probably reflects the evolutionary internalisation of sensory apparatus. Simpler, early animals had external sensory apparatus that allowed the animal to sense nutrients, chemicals, light, and so on. This apparatus developed from ectoderm, the external layer. In humans much of this sensory apparatus remains external to some extent (the retina, touch, temperature and pain senses in the skin), but the nervous system that has evolved is now located internally.

Mesoderm

The mesoderm is a major contributor to the embryo and its cells are used to build the bones, cartilage and connective tissues of the skeleton, striated skeletal muscle, smooth muscle, most of the cardiovascular system and lymphatic system, the reproductive system, kidneys, the suprarenal cortex, ureters, the linings of body cavities such as the peritoneum, the dermis of the skin and the spleen (Figure 14.2). Cells of the cardiovascular and immune systems formed in the bone marrow are also derived from mesoderm.

Endoderm

So the remainder of the embryo must be formed from endoderm. What is left? Epithelia derived from endoderm line internal passages exposed to external substances, including the gastrointestinal tract, the lungs and respiratory tracts (Figure 14.3). Glands that open into the gastrointestinal tract and the glandular cells of organs associated with the gastrointestinal tract, such as the pancreas and liver, are also derived from the endoderm (Figure 14.3).

The epithelia of the urethra and bladder come from endoderm cells, as do the tonsils, the thymus, the thyroid gland and the par-athyroid glands (Figure 14.3). You can find out how these latter structures form in the pharyngeal arch chapters (see Chapters 38–41).

Endoderm and ectoderm meet at the openings of the mouth and anus. Thus, the oral cavity and part of the pharynx have an epithelium derived from ectoderm, and the remainder of the pharynx has a lining derived from endoderm. The same thing occurs at the anus, and this has important anatomical ramifications for the development of the vasculature there, with respect to portosystemic venous anastomoses for example.

Germ cells

The germ layers should not be confused with germ cells. Germ cells migrate from the yolk sac through the gut tube and dorsal mesentery into the dorsal mesenchyme of the embryo (see Chapter 36). Here they differentiate to form gametes; either oocytes or spermatocytes.

A gamete is a reproductive cell with a haploid (half) set of chromosomes that will combine with another gamete during fertilisation to produce a new cell with a full, diploid complement of chromosomes (see Chapter 9). That cell (zygote) will become the embryo and its supporting structures.

A gamete, then, is an example of a cell that does not develop from embryonic ectoderm, endoderm or mesoderm.

Clinical relevance

With the meeting of the ectoderm and endoderm near the anal opening of the gut tube, the rectum develops with links partly to the ectoderm and partly to the endoderm. The superior part of the rectum (endoderm, gut tube) drains blood back to the liver via the superior rectal vein and subsequently the inferior mesenteric vein and the portal vein. The inferior part of the rectum (ectoderm, not gut tube) drains blood via the inferior and middle rectal veins to pelvic veins, iliac veins, the inferior vena cava and thus into the systemic circulation. Venous anastomoses (links) exist between the superior, middle and inferior rectal veins. An impedance to the flow of blood through the liver from the portal vein to the inferior vena cava will cause the blood to find an alternate route, one example of which are the rectal **portosystemic venous anastomoses**. The veins here stretch and enlarge, causing haemorrhoids.

Germ cell tumours are growths that develop from germ cells. They may occur within or outside the gonads, possibly from aberrant or normal migration, and can be congenital.

Germ cell tumours of different types exist, including teratomas. A **teratoma** may form structures of any of the three germ *layers*, including thyroid, liver or lung tissues, or occasionally hair, teeth or bone.

Figure 15.1
Early neurulation

Figure 15.2
Signals from the notochord start off the processes of neurulation in the ectoderm

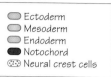

Ectoderm
Mesoderm
Endoderm
Notochord
Neural crest cells

Involution

Figure 15.3
Ectoderm involutes, starts to form a tube

Figure 15.4
Neural crest cells appear in the crests of the waves of the ectoderm that are moving towards each other

Neural crest cells migrate

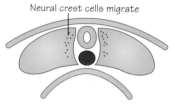

Figure 15.5
The neural tube has formed, and neural crest cells move away

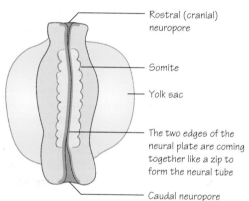

Rostral (cranial) neuropore

Somite

Yolk sac

The two edges of the neural plate are coming together like a zip to form the neural tube

Caudal neuropore

Figure 15.6
Embryo around day 22–23. The neural tube has formed but neuropores have yet to close

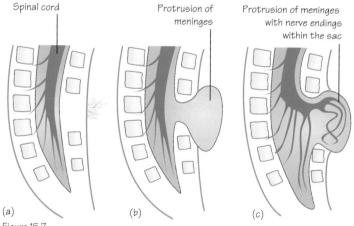

Spinal cord

Protrusion of meninges

Protrusion of meninges with nerve endings within the sac

(a)   (b)   (c)

Figure 15.7
Classifications of spina bifida: (a) spina bifida occulta, (b) meningocele and (c) myelomeningocele (also known as cystica)

Embryology at a Glance, First Edition. Samuel Webster and Rhiannon de Wreede.

38 © 2012 John Wiley & Sons, Ltd. Published 2012 by John Wiley & Sons, Ltd.

Time period: days 18–28

Introduction

The formation of the neural tube from a flat sheet of ectoderm is called **neurulation**. The initially simple tube will develop and form the brain, spinal cord and retina, and is the source of neural crest cells and their derivatives.

Notochord

As cells of the epiblast pass through the primitive streak during gastrulation, some of those cells are destined to form a distinct collection of cells in the midline of the developing embryo.

The **primitive node** extends as a tube of mesenchymal cells running in the midline of the embryo between the ectoderm and endoderm. This is the **notochordal process**. It grows and extends in a cranial direction developing a lumen.

Around day 20 the notochordal process fuses with the endoderm beneath it, forming the **notochordal plate**. A couple of days later the cells of the notochordal plate lift from the endoderm and form a solid rod, again running almost the full length of the midline of the embryo. This is the **notochord** (Figure 15.1).

Neural plate

The notochord is a signalling centre that signals to the cells of the overlying ectoderm. As the notochord forms the ectoderm in the midline of the embryo thickens, becoming the **neural plate** from day 18 (Figure 15.2). Now the ectoderm is becoming neuroectoderm. This begins at the cranial end of the embryo and extends towards the caudal end.

The neural plate is broader cranially, and this will form the brain. The remainder of the neural plate elongates and develops into the spinal cord.

Neural tube

The neural plate dips inwards in the midline, beginning to fold and form a **neural groove** (Figure 15.3). The sides of the groove are the **neural folds**, and the parts of neuroectoderm brought towards one another to meet are the **neural crests**. The neural crests look like the crests of two waves crashing into each other to complete the tube.

The two sides of the neural plate are brought together, meet and fuse, forming a self-contained tube of neuroectoderm running the length of the embryo, open at either end (Figure 15.4). This is the **neural tube**.

The neural tube separates from the ectoderm, which reforms over the neural tube, forming the external surface of the embryo (Figure 15.5).

Development of the neural tube from the neural plate extends cranially and caudally, leaving either end open at the cranial and caudal **neuropores** (Figure 15.6). The cranial neuropore closes on day 24 and the caudal neuropore closes on day 26. Neurulation is now complete.

Neural crest cells

As the neural tube forms from the neural plate a new cell type appears in the neural crest. These are **neural crest cells** (Figure 15.4), and as the neural tube forms these cells leave the neural tube and migrate away to other parts of the embryo (Figure 15.5). They become parts of a wide range of organs and structures, and differentiate to form a variety of different cell types.

For example, they will form much of the peripheral nervous system, skeletal parts of the face and pigment cells in the skin (melanocytes). Migration and differentiation of these cells is well organised and an important part of the normal development of much of the embryo.

Development of the central nervous system

From neurulation the central nervous system continues to develop as the cranial end of the neural tube dilates and folds to form spaces that will become the brain. The remainder of the neural tube, caudal to the first 4 somites, will become the spinal cord.

Cells of the walls of the tube differentiate and proliferate to become neurons, glial cells and macroglial cells, and the walls thicken. You can read about the development of the central nervous system in Chapter 42.

Clinical relevance

The most common congenital abnormalities of neurulation are neural tube defects. As the neuropores are the last parts of the neural tube to close, defects are most likely to occur at its cranial or caudal ends.

Failure of the neural tube to close caudally affects the spinal cord and the tissues that overlie it, including the meninges, vertebral bones, muscles and skin.

Spina bifida (from the Latin for 'split spine') is a condition in which vertebrae fail to form completely. It may manifest in different degrees of severity. **Spina bifida occulta** is the least severe form with a small gap in one or more vertebrae in the region of L5–S1 (Figure 15.7), often causing little or no symptoms. An unusual tuft of hair may be present in this region of the back.

Spina bifida meningocoele is a failure of vertebrae to fuse that is large enough to allow the protrusion of the meninges of the spinal cord externally (Figure 15.7). If the spinal cord or nerve roots also protrude this is called spina bifida with meningomyelcoele. This may affect sensory and motor innervation at the level of the lesion, potentially affecting bladder and anal continence.

The neural tube may also fail to close at the cranial end, causing abnormal brain and calvarial bone development. The brain may be partly outside the skull (**exencephaly**) or the forebrain may fail to develop entirely (**anencephaly**). Exencephaly may precede anencephaly as the extruded brain tissue degenerates. Anencephaly is incompatible with life.

The incidence of neural tube defects is reduced by folic acid supplements in the diet, but as neurulation occurs during the third and fourth weeks it should be considered early in pregnancy or when trying to conceive.

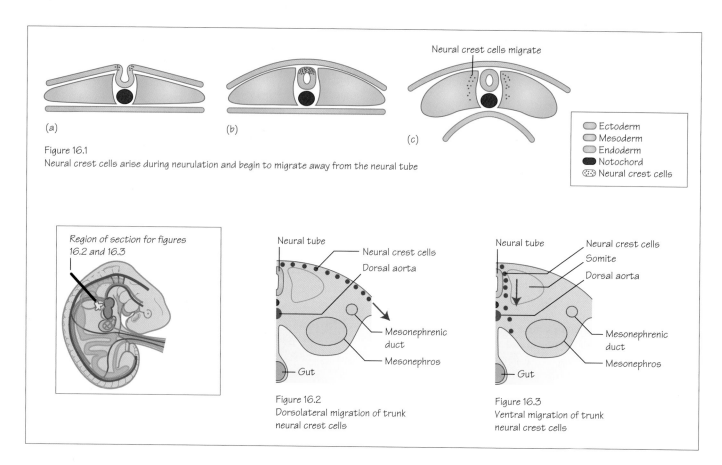

(a)

(b)

(c)

Neural crest cells migrate

Ectoderm
Mesoderm
Endoderm
Notochord
Neural crest cells

Figure 16.1
Neural crest cells arise during neurulation and begin to migrate away from the neural tube

Region of section for figures 16.2 and 16.3

Neural tube
Neural crest cells
Dorsal aorta

Mesonephric duct
Mesonephros
Gut

Figure 16.2
Dorsolateral migration of trunk neural crest cells

Neural tube
Neural crest cells
Somite
Dorsal aorta

Mesonephric duct
Mesonephros
Gut

Figure 16.3
Ventral migration of trunk neural crest cells

Embryology at a Glance, First Edition. Samuel Webster and Rhiannon de Wreede.

Time period: from day 22

Neural crest cells

During neurulation (see Chapter 15) a group of cells arises in the crests of the neural plates that are brought together to form the neural tube (Figure 16.1). These neural crest cells migrate out of and away from the neural tube to other parts of the developing embryo. As they break cell contacts and leave the neuroectoderm they become mesenchymal. The term mesenchyme typically refers to the connective tissue of the embryo formed from the mesoderm. Neural crest cells become histologically similar to the cells of the mesenchyme.

They migrate, proliferate and differentiate into a number of different adult cell types, contributing to many structures and organs, and you will find them throughout this book. As they are able to differentiate into a number of different cell types they are regarded as **multipotent** rather than **pluripotent**, like many of the cells of the embryo at this stage.

Migration and differentiation

The migration of neural crest cells begins in the cranial end of the embryo shortly before the neuropores of the neural tube close. Although they soon become interspersed amongst the cells of the embryo that they are moving through, they can be tracked in the lab with cell labelling techniques.

A cranial group of neural crest cells migrates dorsolaterally to take part in formation of structures of the head and neck. Two groups of trunk neural crest cells migrate in different directions; either dorsolaterally around towards the midline ectoderm (Figure 16.2) or ventrally around the neural tube and notochord (Figure 16.3).

When migrating neural crest cells encounter an obstacle that prevents further progress they tend to clump and accumulate. An obstacle may be another group of cells, a basal lamina or extracellular matrix molecules such as chondroitin sulphate-rich proteoglycans. A barrier to migration may cause the neural crest cells to migrate along it in a particular direction. Other extracellular matrix molecules such as fibronectin, proteoglycans and collagen will also affect the migration of neural crest cells. By altering the localisation and concentration of molecules that aid, encourage or inhibit migration the final location of neural crest cells can be modified by the embryo.

Destinations

Differentiation of neural crest cells occurs in response to a range of external stimuli encountered during migration.

Neural crest cells taking the dorsolateral routes towards the ectoderm of the embryo will differentiate into the melanocytes of the skin, for example. Some neural crest cells in the trunk region that migrate ventrally will become neurons of the dorsal root ganglia and sympathetic ganglia (see Chapters 37 and 43).

Neural crest cell derivatives

- Melanocytes (skin)
- Dermis, some adipose tissue and smooth muscle of the neck and face (skin)
- Neurons (dorsal root ganglia)
- Neurons (sympathetic ganglia)
- Neurons (ciliary ganglion)
- Neurons (cranial sensory V, VII, maybe VIII, IX, X)
- Schwann cells (nervous system)
- Adrenomedullary cells (adrenal glands)
- Enteric nervous system (gastrointestinal tract, parasympathetic nervous system)
- Craniofacial cartilage and bones (musculoskeletal)
- Bones of the middle ear (musculoskeletal)
- Thymus (immune system)
- Odontoblasts (teeth)
- Conotruncal septum (heart)
- Semilunar valves (heart)
- Connective tissue and smooth muscle of the great arteries (aorta, pulmonary trunk)
- Neuroglial cells (central nervous system)
- Parafollicular cells (thyroid gland)
- Glomus type I cells (carotid body)
- Connective tissue of various glands (salivary, thymus, thyroid, pituitary, lacrimal glands)
- Corneal endothelium, stroma (eye)

Clinical relevance

Neural crest cells are obviously important in various areas of embryological development, and they must migrate in a very organised manner to complete this development normally.

Sometimes, neural crest cells do not migrate to their intended destinations. For example, a deficiency in the number of neural crest cells available to form mesenchyme in the developing face can cause **cleft lip** and **cleft palate**.

Albinism may be caused by a failure of neural crest cell migration but is more likely to be caused by a defect in the melanin production mechanism. However, pigmentation anomalies are apparent in patients with **Waardenburg syndrome**, such as eyes of different colours, a patch of white hair or patches of hypopigmentation of skin. Waardenburg syndrome is associated with an increased risk of hearing loss, facial features such as a broad, high nasal root and cleft lip or palate. Gene mutations of one of at least four genes can cause Waardenburg syndrome, including *Pax3*, a gene involved in controlling neural crest cell differentiation.

An abnormality of migration of neural crest cells into the pharyngeal arches can lead to improper development of the parathyroid glands, thymus, facial skeleton, heart, aorta and pulmonary trunk. This is 22q11.2 deletion syndrome or **DiGeorge syndrome** (also known as CATCH22 syndrome). Congenital defects vary between patients with DiGeorge syndrome but it is likely that they will suffer hypocalcaemia, a cleft palate, a conotruncal defect such as a ventricular septal defect or tetralogy of Fallot, recurrent infections, renal problems and learning difficulties. These varied structures are linked by their development from neural crest cells and pharyngeal arches.

17 Body cavities (embryonic)

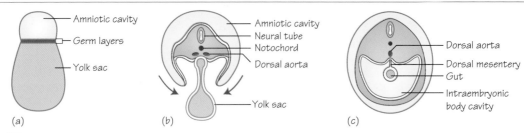

- Amniotic cavity
- Germ layers
- Yolk sac

(a)

- Amniotic cavity
- Neural tube
- Notochord
- Dorsal aorta
- Yolk sac

(b)

- Dorsal aorta
- Dorsal mesentery
- Gut
- Intraembryonic body cavity

(c)

Figure 17.1
Formation of the intraembryonic cavity, (a) location of the germ layers within the embryo, (b) movement of the embryo pinches off the yolk sac, (c) formation of the intraembryonic cavity in week 4

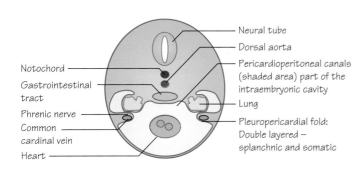

- Neural tube
- Dorsal aorta
- Pericardioperitoneal canals (shaded area) part of the intraembryonic cavity
- Lung
- Pleuropericardial fold: Double layered – splanchnic and somatic

- Notochord
- Gastrointestinal tract
- Phrenic nerve
- Common cardinal vein
- Heart

Figure 17.2
Development of the pleuropericardial folds and the pericardioperitoneal canals at approximately 9 weeks gestation

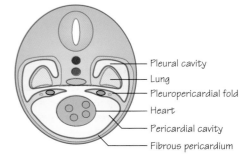

- Pleural cavity
- Lung
- Pleuropericardial fold
- Heart
- Pericardial cavity
- Fibrous pericardium

Figure 17.3
Development of the thoracic cavity, formation of the pleural and pericardial cavities. As the lungs grow more anteriorly, the pleuropericardial folds fuse with each other and with the root of the lungs. The phrenic nerve ends up residing within the fibrous pericardium, which completely surrounds the heart

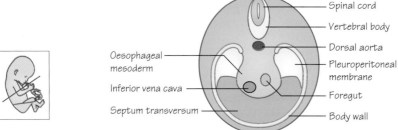

- Spinal cord
- Vertebral body
- Dorsal aorta
- Pleuroperitoneal membrane
- Foregut
- Body wall

- Oesophageal mesoderm
- Inferior vena cava
- Septum transversum

Figure 17.4
The development of the diaphragm at around week 7. Fusion of the pleuroperitoneal folds with the septum transversum, the oesophageal mesentery and muscular ingrowth from the body walls

Embryology at a Glance, First Edition. Samuel Webster and Rhiannon de Wreede.

Time period: day 21 to week 8

Body cavities

From a tightly packed, flat trilaminar disc of cells the body cavities must form. This is initiated around 21 days in the lateral plate mesoderm, which splits into splanchnic and somatic divisions. Between these mesodermal divisions vacuoles form and merge creating a U-shaped cavity in the embryo. This is the **intra-embryonic cavity** (Figure 17.1) and initially has open communication with the **extra-embryonic cavity** (or chorionic cavity).

When the embryo folds the connection with the chorionic cavity is lost resulting in a cavity from the pelvic region to the thoracic region of the embryo.

Of the two layers of lateral plate mesoderm that divided, a somatic layer lines the intra-embryonic cavity and a splanchnic layer covers the viscera.

The septum transversum divides the cavity into two: the **thoracic** and **abdominal (peritoneal)** cavities. The division is not complete and there remains communication between these cavities through the **pericardioperitoneal canals** (Figure 17.2).

Membranes develop at either end of these canals. These membranes separate the thoracic cavity into the pericardial cavity and pleural cavities and are called **pleuropericardial folds** (Figures 17.2 and 17.3). The folds carry the phrenic nerves and common cardinal veins and as the position of the heart changes inferiorly, the folds fuse. The pleuropericardial folds will form the fibrous pericardium (Figure 17.3).

Diaphragm

The diaphragm consists of components of the septum transversum, pleuroperitoneal folds, some oesophageal mesentery and a little muscular ingrowth from the dorsal and lateral body walls (Figure 17.4).

The **septum transversum** originates around day 22 at a cervical level, but caudal to the developing heart. It receives innervation from spinal nerves C3–C5, the beginning of the **phrenic nerve**. With growth of the embryo the position alters to rest at the level of the thoracic vertebrae.

The septum transverum is a boundary between the abdominal cavity and the thoracic cavity. There are two connections between these cavities as mentioned above; the pericardioperitoneal canals.

The **pleuroperitoneal folds** arise from the dorsal body wall and eventually close off the pericardioperitoneal canals and prevent communication between the abdominal and thoracic cavities.

The pleuroperitoneal folds fuse with the septum transversum, the oesophageal mesentery and the muscular ingrowth from the body walls to form the diaphragm. Muscle cells from the septum transversum and the body wall invade the folds forming the muscular part of the diaphragm (Figure 17.4). The septum transversum forms the central tendon and the mesentery of the oesophagus merges into the central tendon, thus allowing passage of the aorta, vena cava and oesophagus.

Clinical relevance

In a congenital diaphragmatic hernia, caused by a failure of the diaphragm to form completely, the abdominal contents herniate into the thoracic cavity negatively affecting lung development, leading to pulmonary hypoplasia and hypertension. Generally survival rates are about 50%, but if the liver is unaffected they are nearer 90%. Treatment involves mechanical ventilation and extracorporeal membrane oxygenation (ECMO) to perform gas exchange, and even a lung transplant has been successfully reported.

Gastroschisis is also a herniation of the bowel, but caused by an anterior abdominal wall defect, usually just to the right of the umbilicus. Viscera are not covered with peritoneum or amnion, and it is not associated with the same level of other abnormalities (unlike omphalocoele). Surgical intervention is required and generally survival rates are good.

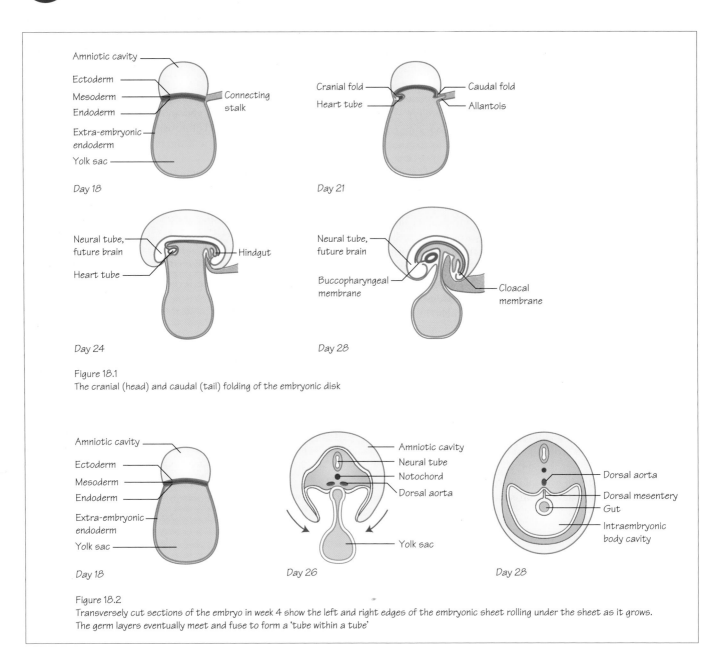

Amniotic cavity
Ectoderm
Mesoderm
Endoderm
Extra-embryonic endoderm
Yolk sac
Connecting stalk

Day 18

Cranial fold
Heart tube
Caudal fold
Allantois

Day 21

Neural tube, future brain
Heart tube
Hindgut

Day 24

Neural tube, future brain
Buccopharyngeal membrane
Cloacal membrane

Day 28

Figure 18.1
The cranial (head) and caudal (tail) folding of the embryonic disk

Amniotic cavity
Ectoderm
Mesoderm
Endoderm
Extra-embryonic endoderm
Yolk sac

Day 18

Amniotic cavity
Neural tube
Notochord
Dorsal aorta
Yolk sac

Day 26

Dorsal aorta
Dorsal mesentery
Gut
Intraembryonic body cavity

Day 28

Figure 18.2
Transversely cut sections of the embryo in week 4 show the left and right edges of the embryonic sheet rolling under the sheet as it grows. The germ layers eventually meet and fuse to form a 'tube within a tube'

Embryology at a Glance, First Edition. Samuel Webster and Rhiannon de Wreede.

Time period: days 17–30

Flat sheet
After the formation of the three germ layers of the embryo during week 3 (see Chapter 13), the embryo remains a flat, oval sheet of cells with an amniotic cavity above it and a yolk sac beneath. Differential growth of these embryonic and extra-embryonic cells causes the flat embryo to curve and fold at the head end, the tail end and laterally. With this folding and rolling up the embryo begins to take on the early shape of a body.

Longitudinal folding
As the flat embryo grows, its amniotic cavity grows, but the yolk sac does not. The enlarging sheet of the embryo pushes out and over the rim of the yolk sac, and is pulled around and underneath itself (Figure 18.1).

As the cranial fold progresses, the buccopharyngeal (or oropharyngeal) membrane (see Figure 13.2) moves around to the position of the future mouth, and the early neural tube that will form the brain comes to lie cranially to it. A region of cells that begin to form the heart tube (see Chapter 25) are also pulled around and come to lie in the future thorax, caudal to the mouth.

At the caudal end, folding brings the cloacal membrane (Figure 13.2) underneath the embryo, and the connecting stalk around towards the future umbilical region of the embryo's abdomen. With this movement the connecting stalk, the allantois and the yolk sac are all brought close together (Figure 18.1). The connecting stalk is the link between the embryo and the placenta. The yolk sac by this stage (day 26) is linked to the early gastrointestinal tract by the vitelline duct (see Chapter 31).

Lateral folding
As the embryo curls up longitudinally, it also rolls up across its width. The left and right flanks of the embryonic disc extend and curl around underneath the embryo, squeezing the sides of the yolk sac (see Figure 17.1).

The left and right flanks meet, and the germ layers of either side meet and fuse. The ectoderm of the left side meets the ectoderm of the right side forming a continuous external surface for the embryo. Similarly, the mesodermal and endodermal layers meet. The endoderm forms a tube that ends at the buccopharyngeal and cloacal membranes, which also remains continuous with the yolk sac. This is the lining of the gastrointestinal tract (see Chapter 31). This meeting of the left and right flanks or folds of the embryo begins at the cranial and caudal ends and continues towards the middle. By day 30 the yolks sac's connection to the gastrointestinal tract is squeezed by this growth, but remains substantial (Figure 18.1).

Tube within a tube
As a result of this folding, curving, rolling and pinching, the embryo has a 'tube within a tube' body plan at the start of week 5. The outer tube is made of ectoderm, the inner tube is endoderm, and in between lies mesoderm and the early body cavity (also known as the coelom). This arrangement is common to many embryos, from nematodes to humans, and marks a major trend in evolution.

Clinical relevance
Gastroschisis describes the herniation of abdominal contents externally through the anterior abdominal wall. It is usually detected before birth by ultrasound, and the defect often lies to one side of the umbilicus. Gastroschisis may result from a failure of the anterior body wall to form normally as described above. It can be treated after birth surgically or by protecting the herniated bowel in an aseptic film and allowing the intestine to return to the abdominal cavity slowly over time. Omphalocoele is a different type of foetal herniation, in which the abdominal contents herniate into the umbilicus and are therefore covered (see Chapter 33).

Head   Tail

Figure 19.1
The early Drosophila embryo has a striped pattern of gap gene expression

Figure 19.2
Pair rule genes are expressed in alternating stripes by the cells of the embryo, and segments can be visualised by looking for this pattern

Figure 19.3
Segment polarity genes are expressed in bands within the segments

Drosophila embryo   Drosophila adult

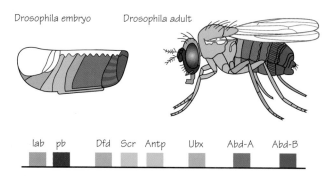

lab   pb   Dfd   Scr   Antp   Ubx   Abd-A   Abd-B

Figure 19.4
Hox genes begin the specification of segments of the embryo for morphogenesis to form different structures, (e.g. legs or antennae)

Figure 19.5
If Hox gene expression is disrupted segments are not specified correctly, and can instead develop like a different segment. In the case of the Drosophila antennapedia mutant here, the fly develops legs where it would normally have antennae

Tail (growing)

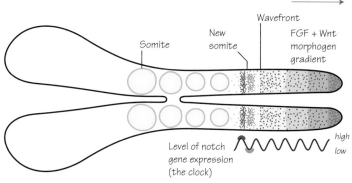

Somite   New somite   Wavefront   FGF + Wnt morphogen gradient

Level of notch gene expression (the clock)   high   low

Figure 19.6.
The segmentation clock
Cells in the presomitic mesoderm oscillate from high to low levels of expression of genes of the Notch pathways. As these cells leave the morphogen gradient they leave at a point of high (blue) or low (green) levels of Notch expression. This determines their formation of cranial or caudal parts of the somite

Embryology at a Glance, First Edition. Samuel Webster and Rhiannon de Wreede.

Time period: days 18–35

Introduction

Segmentation is an important concept in embryology. Early animals, for example nematodes or very early insects, are built around a repeating pattern. The segments of later insects are also repeated, but some have become specialised with modified legs, mouth parts or wings. The evolution of these changes is recorded, to an extent, within the genes responsible for early organisation and patterning of the embryos of these animals.

Pair rule genes

A common insect used for investigating and discussing embryology, and segmentation in particular, is the fruit fly, also known as *Drosophila melanogaster*.

The cells of the *Drosophila* embryo are initially organised along a craniocaudal axis by a morphogen gradient (see Chapter 3 for a similar example of a morphogen gradient). This is followed by the expression of different genes by the cells of the embryo, but only in particular bands along its length. These are **gap genes** (Figure 19.1).

This banded pattern of gene expression becomes more pronounced when **pair rule genes** are expressed in alternating stripes by the cells of the embryo (Figure 19.2). This level of organisation is pushed even further by the expression of **segment polarity genes** within those segments (Figure 19.3).

Hox genes

Now that the embryo is organised into similar segments, the cells of each segment need further information from which morphogenesis will shape the appropriate structures for each segment (e.g. a wing, or a leg).

Hox genes are genes that share a similar **homeobox** domain of 180 base pairs, which encodes for a sequence of 60 amino acids. The term 'homeobox' refers to the sequence of base pairs, and the term 'homeodomain' refers to the section of protein that corresponds to the homeobox. The homeodomain is highly conserved between genes and between species, with small differences.

Hox genes are involved in the very early specification of the segments of the embryo, from which the development of morphologically different segments can occur. They are expressed in bands along the length of the embryo (Figure 19.4), and in vertebrates there are multiple, overlapping, similar sets of Hox genes (clusters) that gives some redundancy and more complex organisation than possible in the development of the fly. The Hox genes of *Drosophila* do not have this redundancy, so knocking out Hox genes gives profound effects. A common example is the Antennapedia mutant, in which the fly develops legs where its antennae would normally form (Figure 19.5). The Hox gene that would normally specify this segment is lost, the pattern is broken and the segment is re-specified.

Hox genes are found together on the same chromosome, lined up. Interestingly, they are lined up in their order of expression along the craniocaudal axis. In humans the 4 clusters of Hox genes are found on 4 different chromosomes.

Hox proteins

The Hox proteins that result from Hox gene expression are DNA binding transcription factors, able to switch on cascades of genes. The homeodomain is the DNA binding region of the protein.

Segmentation clock

All of this organisation leads to the formation of visible early segmentation patterns such as the somites (see Chapter 20), from which adult segmented structures develop. In humans and other vertebrates the segmentation pattern can be seen in the vertebrae, ribs, muscles and nervous innervation patterns (see Figure 20.5). These segments form sequentially, one pair after another.

Before somites form, cells of the presomitic mesoderm display oscillating patterns of gene expression, meaning the expression of genes switches on, off and on again with time. This rhythmic expression of genes of the Notch pathways and their targets is known as the segmentation clock. You can think of each cell having its own clock and its own time.

A morphogen gradient of fibroblast growth factor (FGF) and Wnt is secreted by cells at the tail end of the presomitic mesoderm. You might call the edge of this morphogen gradient the wavefront.

As cells at the caudal end of the presomitic mesoderm proliferate and the tail grows, the cells producing FGF and Wnt move further away from the head and from other presomitic mesoderm cells. Some cells of the presomitic mesoderm no longer feel the effects of FGF and Wnt as the wavefront moves away from them, and they begin to form somites.

The band of cells that leave the wavefront will either form the cranial end or the caudal end of the somite depending upon the time of their segmentation clock at the point at which they leave the wavefront. The temporal nature of the segmentation clock is translated into the spatial arrangement of somites via these mechanisms (Figure 19.6).

How the segmentation clock works is still not entirely understood, but the understanding of these mechanisms has developed remarkably over the last 15 years.

Vertebrates

Through the embryology of segmentation we can see the path of evolution and links between vastly different animals, existing now and in prehistory, and the mechanisms behind anatomical similarities amongst vertebrates. The giraffe, for example, has 7 cervical vertebrae in its very long neck, just as we do in our much shorter variant. While segmentation is clearly apparent in the bony structures of adult anatomy, the embryology here helps us understand the arrangement of many of the soft tissues too.

Clinical relevance

Minor errors in segmentation can produce vertebral and intervertebral defects. A wedge-shaped **hemivertebra** may form, causing a form of **congenital scoliosis** that worsens as the hemivertebra grows. A number of variations have been documented. Other vertebrae may be fused completely, just laterally, posteriorly or anteriorly, if the intervertebral space fails to form completely causing **kyphosis** or **lordosis**. Other developmental processes may also cause these deformities.

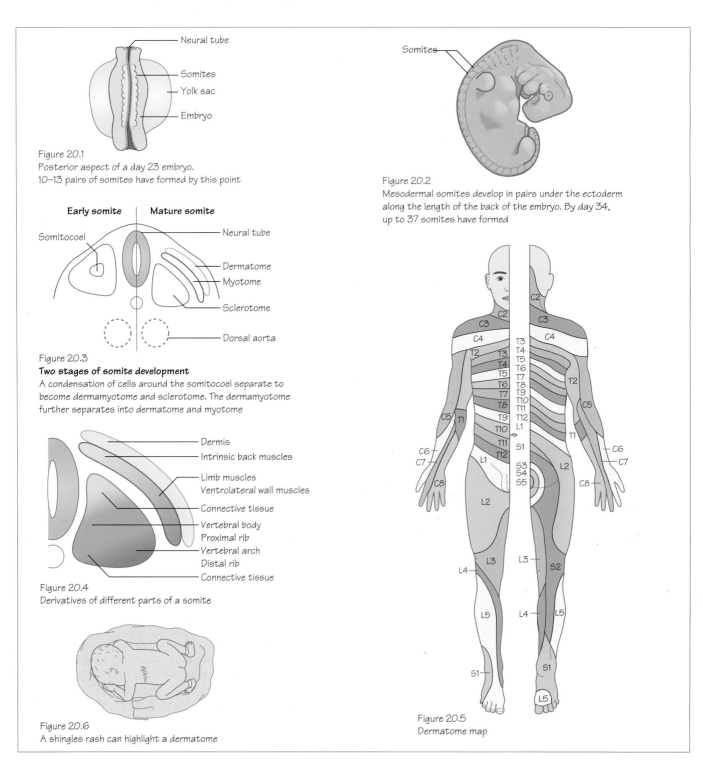

Figure 20.1
Posterior aspect of a day 23 embryo.
10–13 pairs of somites have formed by this point

Figure 20.2
Mesodermal somites develop in pairs under the ectoderm along the length of the back of the embryo. By day 34, up to 37 somites have formed

Figure 20.3
Two stages of somite development
A condensation of cells around the somitocoel separate to become dermamyotome and sclerotome. The dermamyotome further separates into dermatome and myotome

Figure 20.4
Derivatives of different parts of a somite

Figure 20.5
Dermatome map

Figure 20.6
A shingles rash can highlight a dermatome

Time period: days 20–35

Mesoderm

In the formation of the trilaminar disc we see the 3 layers of the cells of the embryo becoming organised as ectoderm, mesoderm and endoderm (see Chapter 14). The mesoderm layer is further organised into areas of **paraxial mesoderm** (medially), **intermediate mesoderm** and **lateral mesoderm** (laterally). These areas of mesoderm will contribute to the formation of different structures (see Figure 23.1).

Embryology at a Glance, First Edition. Samuel Webster and Rhiannon de Wreede.

The somite

A clumping of cells and a thickening of the mesodermal layer on either side of the midline of the embryo forms from paraxial mesoderm and gives the first pair of **somitomeres**. Here we see the beginning of the characteristic segmentation of vertebrate animals. In the cranial region, the first 7 somitomeres contribute to the development of the musculature of the head, but the remaining somitomeres become **somites**.

Somites are cuboidal-shaped condensations (groupings) of cells visible upon the surface of the embryo (Figures 20.1 and 20.2). The organisation of cells here will give rise to much of the axial musculoskeletal system and body wall of the embryo.

What signals initiate somite formation? The answer to this is complex, but many signals come from the overlying ectoderm. Notch signalling and Hox genes are certainly involved here, amongst others (see Chapter 19).

The first somite forms during day 20 and subsequent somites appear at a rate of 3 pairs a day. Somites form in a cranial to caudal sequence, lying laterally to the neural tube. By the end of week 5 a full complement of somites will have formed, including 4 occipital, 8 cervical, 12 thoracic, 5 lumbar, 5 sacral and 8–10 coccygeal pairs. The number of visible somites is often used as a method of dating (or staging) an embryo.

The first occipital and last 5–7 coccygeal somites degenerate, so from the 42–44 pairs of somites that form around 37 remain. The tightly packed cells of the remaining somites develop a lumen in their centres, termed the **somitocoele** (Figure 20.3). The somitocoele cells are involved in many complex interactions resulting in **epithelialisation** (layering) and polarity within the somite (cells become organised).

The cells in each somite differentiate and move to give ventral and dorsal groups of cells called the **sclerotome** and **dermomyotome**, respectively (Figure 20.3).

Sclerotome

The cells of the ventromedial part of the somite form the sclerotome. When they lose their tight bindings to one another they migrate to surround the notochord.

These cells will form the vertebrae, the intervertebral discs, the ribs and connective tissues (Figure 20.4). The caudal part of the sclerotome of one somite and the dorsal part of its neighbouring somite's sclerotome combine to form a single vertebral bone (see Chapter 22).

The word sclerotome is formed from the Greek words *skleros*, meaning 'hard', and *tome*, meaning 'a cutting'. Cells from the sclerotome form hard structures of the axial skeleton.

A specific dorsolateral region in the sclerotome has relatively recently been shown to form the origins of tendons, termed the **syndetome** (see Chapter 23).

Myotome

The dermomyotome mass of cells in the dorsolateral part of the somite splits again into 2 more groups: the myotome and the dermotome (Figure 20.3). The cells of the myotome will become myoblasts and form the skeletal muscle of the body wall.

Medially positioned cells within the myotome form the **epaxial** muscles intrinsic to the back (e.g. erector spinae). Lateral cells will form the **hypaxial** muscles (the muscles of the ventrolateral body wall such as the intercostal muscles and the abdominal oblique and transverse muscles). Laterally placed cells will also migrate out to the limb buds and form the musculature of the limbs (Figure 20.4).

This is covered in a little more detail in Chapters 23 and 24.

Dermotome

The other part of the dermomyotome, the dermotome, is the most dorsal group of cells within the somite. These cells will contribute to the dermis and subcutaneous tissue of the skin of the neck and trunk (Figure 20.4).

Skin

The integumentary system receives contributions from a variety of sources. The epidermis, nails, hair and glands develop from ectoderm, the dermis (connective tissue and blood vessels) develop from mesoderm and the dermotome, and pigmented cells (melanocytes) differentiate from migrating neural crest cells.

Innervation

It is important to note that cell groups retain their innervation from their segment of origin, no matter where the migrating cells end up. A spinal nerve develops at the level of each somite and will comprise a collection of sensory and motor axons.

The groups of cells within each myotome and dermotome will migrate to their final destinations trailing the axons of these neurons in their paths. In the adult clear patterns of innervation segmentation remain, commonly seen by medical students in **dermatome** maps (Figure 20.5).

Dermatomes

Not to be confused with dermotomes, a dermatome is a region of skin that is predominantly supplied by the sensory component of one spinal nerve (Figure 20.5). The dermatomes are named according to the spinal nerve that supplies them. In diagrams the dermatomes are shown as very specific areas, but in reality there is significant overlap between dermatomes. Although sensation may be affected by nerve damage it may not completely numb the area. Also be aware that the overlap between dermatomes varies for the sensations of temperature, pain and touch.

Clinical relevance

The **varicella zoster** virus that causes chickenpox can lie dormant in dorsal root ganglia after the patient has recovered. Later in life the virus may follow the pathway of a spinal nerve to travel to the skin, causing **shingles** (herpes zoster; Figure 20.6). It manifests visibly as a rash restricted to a single dermatome, amongst other symptoms. Sometimes, starkly delineated rashes show the shape of the dermatome derived from a single somite's *dermotome*.

By testing for a loss of sensation in particular dermatomes your knowledge of somitic embryology can also be used to find clues to help identify the level of spinal cord damage in a patient or to determine whether specific spinal nerves have been injured.

21 Skeletal system (ossification)

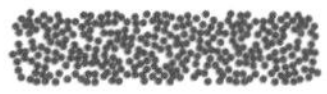

Figure 21.1
Mesenchymal cells condense
and form a model of the future bone

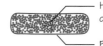
Hypertrophic
chondrocytes

Perichondrium

Figure 21.2
Mesenchymal cells differentiate into
chondrocytes, and the matrix becomes
calcified in the future diaphysis

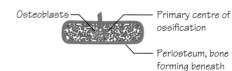
Osteoblasts

Primary centre of
ossification

Periosteum, bone
forming beneath

Figure 21.3
Blood vessels invade, bringing progenitor cells that
become osteoblasts and haematopoietic cells

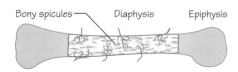
Bony spicules   Diaphysis   Epiphysis

Figure 21.4
The diaphysis becomes ossified but the epiphyses
remain cartilaginous

Secondary
centre of
ossification

Figure 21.5
Later, the epiphyses also begin to ossify

Figure 21.7
Mesenchymal cells form a condensation between 2 developing bones

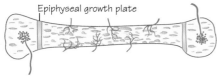
Epiphyseal growth plate

Figure 21.6
With the epiphyses and diaphysis ossified, the bone
continues to grow in length from the growth plates.
Eventually the growth plates also ossify, and growth ceases

Stages of endochondral ossification

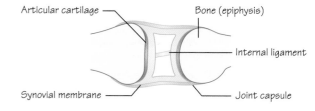
Articular cartilage

Bone (epiphysis)

Internal ligament

Synovial membrane

Joint capsule

Figure 21.8
Mesenchymal cells become organised into layers, and differentiate
into different cell types, in this case the tissues of a synovial joint

Joint development

Embryology at a Glance, First Edition. Samuel Webster and Rhiannon de Wreede.

Time period: week 5 to adult

Introduction

Mesodermal cells form most bones and cartilage. Initially an embryonic, loosely organised connective tissue forms from mesoderm throughout the embryo, referred to as **mesenchyme**. Neural crest cells that migrate into the pharyngeal arches are also involved in the development of bones and other connective tissues in the head and neck (see Chapters 39–42).

Bones begin to form in one of two ways. A collection of mesenchymal cells may group together and become tightly packed (condensed), forming a template for a future bone. This is the start of endochondral ossification (Figure 21.1). Alternatively, an area of mesenchyme may form a hollow sleeve roughly in the shape of the future bone. This is how intramembranous ossification begins.

Long bones form by endochondral ossification (e.g. femur, phalanges) and flat bones form by intramembranous ossification (e.g. parietal bones, mandible).

Endochondral ossification

The cells of the early mesenchymal model of the future bone differentiate to become cartilage (chondrocytes). This cartilage model then begins to ossify from within the diaphysis (the shaft of the long bone). This is the **primary centre of ossification**, and the chondrocytes here enter hypertrophy (Figure 21.2). As they become larger they enable calcification of the surrounding extracellular matrix, and then die by apoptosis.

The layer of perichondrium that surrounded the cartilage model becomes periosteum as the cells here differentiate into osteoblasts, and bone is formed around the edge of the diaphysis. This will become the cortical (compact) bone (Figures 21.2 and 21.3).

Blood vessels invade the diaphysis and bring progenitor cells that will form osteoblasts and haematopoietic cells of the future bone marrow (Figure 21.3). Bone matrix is deposited by the osteoblasts on to the calcified cartilage, and bone formation extends outwards to either end of the long bone (Figure 21.4). Osteoclasts also appear, resorbing and remodelling the new bony spicules of spongy (trabecular) bone.

When osteoblasts become surrounded by bone they are called osteocytes, and connect to one another by long, thin processes through the bony matrix.

The epiphyses (ends) of most long bones remain cartilaginous until the first few years after birth. The **secondary centres of ossification** appear within the epiphyses when the chondrocytes here enter hypertrophy, enable calcification of the matrix and blood vessels invade bringing progenitor cells that differentiate into osteoblasts (Figure 21.5). The entire epiphysis becomes ossified (other than the articular cartilage surface), but a band of cartilage remains between the diaphysis and the epiphysis. This is the **epiphyseal growth plate** (Figure 21.6).

The growth plates contain chondrocytes that continually pass through the endochondral ossification processes described above. A proliferating group of chondrocytes enter hypertrophy in a tightly ordered manner, calcify a layer of cartilage adjacent to the diaphysis, apoptose, and this calcified cartilage is replaced by bone. In this way the long bone continues to lengthen.

Bones grow in width as more bone is laid down under the periosteum. Bone of the medullary cavity is remodelled by osteoclasts and osteoblasts.

When growth ceases at around 18–21 years of age, the epiphyseal growth plates are also replaced by bone (see Chapter 22).

Intramembranous ossification

The flat mesenchymal sleeves that create the templates of flat bones formed by intramembranous ossification contain cells that condense and form osteoblasts directly. Other cells here form capillaries. Osteoblasts secrete a collagen and proteoglycan matrix that binds calcium phosphate, and the matrix (**osteoid**) becomes calcified.

Spicules of bone form and extend out from their initial sites of ossification. Other mesenchymal cells surround the new bone and become the periosteum.

As more bone forms it becomes organised, and layers of compact bone form at the peripheral surfaces (aided by osteoblasts forming under the periosteum), whereas spongy trabeculated bone is constructed in between. Osteoclasts are involved in resorbing and remodelling bone here to give the adult bone shape and structure.

The mesenchymal cells within the spongy bone become bone marrow.

Joint formation

Fibrous, cartilaginous and synovial joints also develop from mesenchyme from 6 weeks onwards. Mesenchyme between bones may differentiate to form a fibrous tissue, as found in the sutures between the flat bones of the skull, or the cells may differentiate into chondrocytes and form a hyaline cartilage, as found between the ribs and the sternum. A fibrocartilage joint may also form, as seen in some midline joints, for example the pubic symphysis.

The synovial joint is a more complex structure, comprising multiple tissues. Mesenchyme between the cartilage condensations of developing limb bones, for example, will differentiate into fibroblastic cells (Figure 21.7). These cells then differentiate further, forming layers of articular cartilage adjacent to the developing bones, and a central area of connective tissue between the bones. The edges of this central connective tissue mass become the **synovial cells** lining the joint cavity (Figure 21.8). The central area degenerates leaving the space of the synovial joint cavity to be filled by synovial fluid. In some joints, such as the knee, the central connective tissue mass also forms **menisci** and **internal joint ligaments** such as the cruciate ligaments.

Clinical relevance

Pregnant women require higher quantities of **calcium** and phosphorus in their diet than normal because of foetal bone and tooth development. Maternal calcium and bone metabolism are significantly affected by the mineralising foetal skeleton, and maternal bone density can drop 3–10% during pregnancy and lactation, and is regained after weaning.

A lack of vitamin D, calcium or phosphorus will cause soft, weak bones to form as the osteoid is unable to calcify. This leads to deformities such as bowed legs and curvature of the spine. Weak bones are more vulnerable to fracture. This is called **rickets**. Other conditions that interfere with the absorption of these vitamins and minerals, or malnutrition during childhood will also lead to rickets. Vitamin D is required for calcium absorption across the gut.

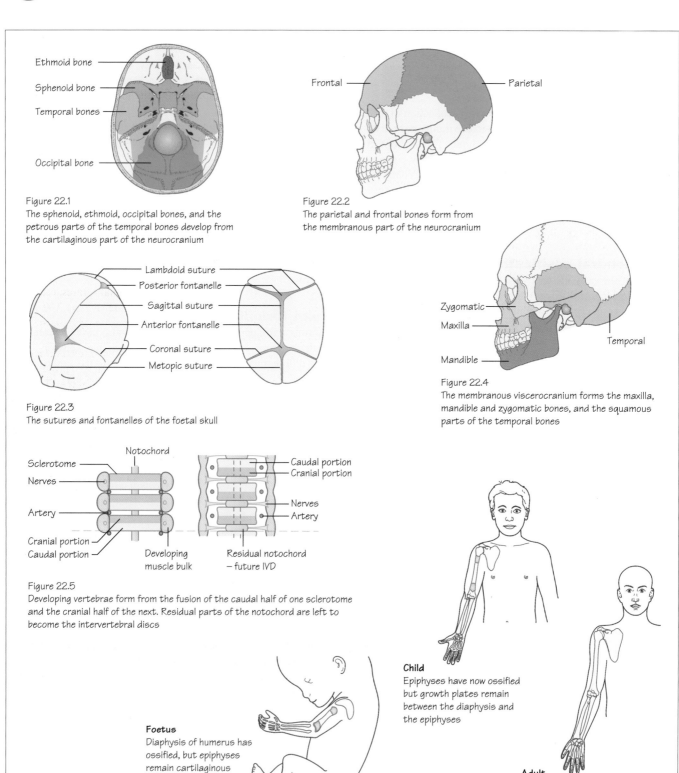

Figure 22.1
The sphenoid, ethmoid, occipital bones, and the petrous parts of the temporal bones develop from the cartilaginous part of the neurocranium

Ethmoid bone
Sphenoid bone
Temporal bones
Occipital bone

Figure 22.2
The parietal and frontal bones form from the membranous part of the neurocranium

Frontal
Parietal

Figure 22.3
The sutures and fontanelles of the foetal skull

Lambdoid suture
Posterior fontanelle
Sagittal suture
Anterior fontanelle
Coronal suture
Metopic suture

Figure 22.4
The membranous viscerocranium forms the maxilla, mandible and zygomatic bones, and the squamous parts of the temporal bones

Zygomatic
Maxilla
Mandible
Temporal

Figure 22.5
Developing vertebrae form from the fusion of the caudal half of one sclerotome and the cranial half of the next. Residual parts of the notochord are left to become the intervertebral discs

Notochord
Sclerotome
Nerves
Artery
Cranial portion
Caudal portion
Developing muscle bulk
Caudal portion
Cranial portion
Nerves
Artery
Residual notochord – future IVD

Figure 22.6.
Ossification of a long bone with age

Foetus
Diaphysis of humerus has ossified, but epiphyses remain cartilaginous

Child
Epiphyses have now ossified but growth plates remain between the diaphysis and the epiphyses

Adult
Growth plates have now ossified

Time period: day 27 to birth

Introduction
Cells for the developing skeleton come from a variety of sources. We have described the development of the somites, and the subdivision of the **sclerotome** (see Chapter 20). Those cells are joined by contributions from the **somatic mesoderm** and migrating **neural crest cells**.

Development of the skeleton can be split into two parts: the **axial** skeleton consisting of the cranium, vertebral column, ribs and sternum; and the **appendicular** skeleton of the limbs.

Cranium
The skull can be divided into another two parts: the **neurocranium** (encasing the brain) and the **viscerocranium** (of the face).

Neurocranium
The bones at the base of the skull begin to develop from cells originating in the occipital somites (paraxial mesoderm) and neural crest cells that surround the developing brain. These cartilaginous plates fuse and ossify (endochondral ossification) forming the sphenoid, ethmoid and occipital bones and the petrous part of the temporal bone (Figure 22.1).

A membranous part originates from the same source and forms the frontal and parietal bones (Figure 22.2). These plates ossify into flat bones (through intramembranous ossification) and are connected by connective tissue sutures.

Where more than two bones meet in the foetal skull a **fontanelle** is present (Figure 22.3). The anterior fontanelle is the most prominent, occurring where the frontal and parietal bones meet. Fontanelles allow considerable movement of the cranial bones, enabling the calvaria (upper cranium) to change shape and pass through the birth canal.

Viscerocranium
Cells responsible for the formation of the facial skeleton originate from the pharyngeal arches (see Chapters 38–41), and the viscerocranium also has cartilaginous and membranous parts during development. The cartilaginous viscerocranium forms the stapes, malleus and incus bones of the middle ear, and the hyoid bone and laryngeal cartilages. The squamous part of the temporal bone (later part of the neurocranium), the maxilla, mandible and zygomatic bones develop from the membranous viscerocranium (Figure 22.4).

Vertebrae
In week 4, cells of the sclerotome migrate to surround the notochord. Undergoing reorganisation they split into cranial and caudal parts (Figure 22.5).

The cranial half contains loosely packed cells, whereas the caudal cells are tightly condensed. The caudal section of one sclerotome joins the cranial section of the next sclerotome. This creates vertebrae that are 'out of phase' with the segmental muscles that reach across the intervertebral joint. When these muscles contract they induce movements of the vertebral column.

Axial bones
Ribs also form from the sclerotome; specifically, the proximal ribs from the ventromedial part and the distal ribs from the ventrolateral part (Figure 20.4). The sternum develops from somatic mesoderm and starts as two separate bands of cartilage that come together and fuse in the midline.

Appendicular bones
Endochondral ossification of the long bones begins at the end of week 7. The primary centre of ossification is the **diaphysis** and by week 12 primary centres of ossification appear in all limb long bones (Figure 22.6).

The beginning of ossification of the long bones marks the end of the embryonic period. Ossification of the diaphysis of most long bones is completed by birth, and secondary centres of ossification appear in the first few years of life within the **epiphyses** (Figure 22.6).

Between the ossified epiphysis and diaphysis the cartilaginous **growth plate** (or epiphyseal plate) remains as a region of continuing endochondral ossification. New bone is laid down here, extending the length of growing bones.

At around 20 years after birth the growth plate also ossifies, allowing no further growth and connecting the diaphysis and epiphysis (Figure 22.6).

Clinical relevance
Cranium
Craniosynostosis is the early closure of cranial sutures, causing an abnormally shaped head. This is a feature of over 100 genetic syndromes including forms of dwarfism. It may also result in underdevelopment of the facial area.

Neural crest cells are often associated with cardiac defects and facial deformations due to failed migration or proliferation. Neural crest cells are also vulnerable to teratogens. Examples of cranial skeletal malformations include: **Treacher Collins** syndrome (mandibulofacial dysotosis), which describes underdeveloped zygomatic bones, mandible and external ears; **Robin sequence** of underdeveloped mandible, cleft palate and posteriorly placed tongue; **DiGeorge** syndrome (small mouth, widely spaced down-slanting eyes, high arched or cleft palate, malar flatness, cupped low-set ears and absent thymus and parathyroid glands).

Vertebrae
Spina bifida is the failure of the vertebral arches to fuse in the lumbosacral region. There are two types. **Spina bifida occulta** affects only the bony vertebrae. The spinal cord remains unaffected but is covered with skin and an isolated patch of hair. This can be treated surgically. **Spina bifida cystica** (meningocoele and myelomeningocoele) occurs with varying degrees of severity. The neural tube fails to close leaving meninges and neural tissue exposed. Surgery is possible in most cases but, because of the increased severity of cystica, continuous follow-up evaluations are necessary and paralysis may occur. It is currently possible to detect spina bifida using ultrasound and foetal blood alpha-fetoprotein levels.

Pregnant women and those trying to become pregnant are advised to take 0.4 mg/day folic acid as it significantly reduces the risk of spina bifida. Folates have an important role in DNA, RNA and protein synthesis.

Scoliosis is a condition of a lateral curvature of the spine that may be caused by fusion of vertebrae, or by malformed vertebrae. The range of treatments for congenital scoliosis includes physiotherapy and surgery. **Klippel–Feil syndrome** is a disease where cervical vertebrae fuse. Common signs include a short neck and restricted movement of the upper spine.

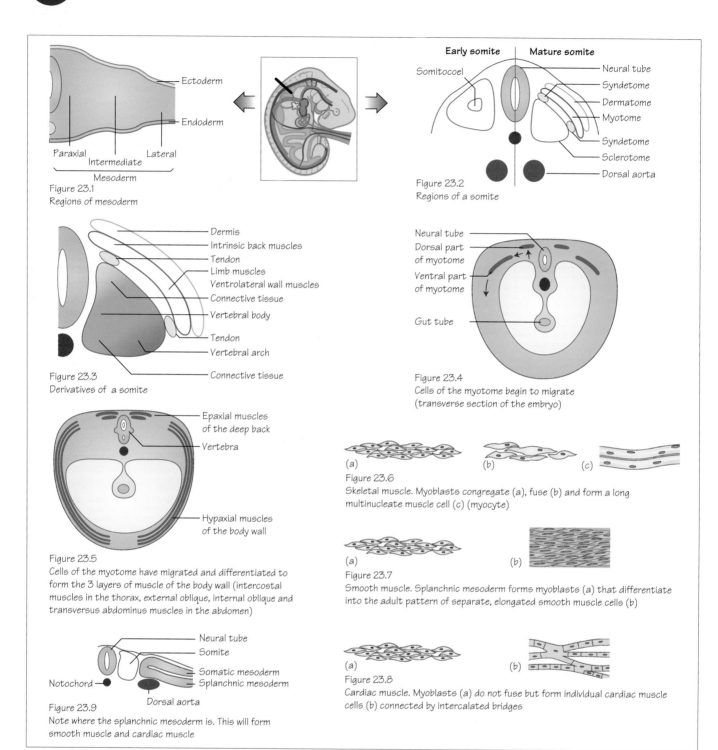

Figure 23.1
Regions of mesoderm

Figure 23.2
Regions of a somite

Figure 23.3
Derivatives of a somite

Figure 23.4
Cells of the myotome begin to migrate
(transverse section of the embryo)

Figure 23.5
Cells of the myotome have migrated and differentiated to
form the 3 layers of muscle of the body wall (intercostal
muscles in the thorax, external oblique, internal oblique and
transversus abdominus muscles in the abdomen)

Figure 23.6
Skeletal muscle. Myoblasts congregate (a), fuse (b) and form a long
multinucleate muscle cell (c) (myocyte)

Figure 23.7
Smooth muscle. Splanchnic mesoderm forms myoblasts (a) that differentiate
into the adult pattern of separate, elongated smooth muscle cells (b)

Figure 23.8
Cardiac muscle. Myoblasts (a) do not fuse but form individual cardiac muscle
cells (b) connected by intercalated bridges

Figure 23.9
Note where the splanchnic mesoderm is. This will form
smooth muscle and cardiac muscle

Embryology at a Glance, First Edition. Samuel Webster and Rhiannon de Wreede.

Time period: day 22 to week 9

Introduction
Most muscle cells originate from the **paraxial mesoderm** (Figure 23.1), and specifically the myotome portion of the somites. The three types of muscle described here are skeletal, smooth and cardiac muscle.

Skeletal muscle
Within each somite the **myotome** splits into two muscle-forming parts: a ventrolateral edge and a dorsomedial edge (Figures 23.2 and 23.3). The ventrolateral edge cells will form the **hypaxial** musculature (i.e. that of the ventral body wall and, in the limb regions, musculature of the limbs) (Figures 23.4 and 23.5). The dorsomedial edge will form the **epaxial** musculature (the back muscles).

During formation of skeletal muscle multiple **myoblasts** (muscle precursor cells) fuse to form myotubes at first, and then long multinucleated muscle fibres (Figure 23.6). By the end of month 3, microfibrils have formed and the striations of actin and myosin patterning associated with skeletal muscle are visible. Important genes involved in myogenesis include *MyoD* and *Myf5*, which cause mesodermal cells to begin to differentiate into myoblasts, and then *MRF4* and *Myogenin* later in the process.

A fourth part of the somite, the **syndetome**, has been recently shown to contain precursor cells of tendons (Figures 23.2 and 23.3). The cells of the syndetome lie at the ventral and dorsal edges of the somites between the cells of the myotome and sclerotome; blocks of cells whose tissues they will connect. They also migrate, but develop independently of muscles and connect later in development. However, tendon cells will also arise from lateral plate mesoderm to populate the limbs, so the full story of tendon development is not limited to the somite.

Limbs
The upper limb bud is visible from day 26 around the levels of cervical somite 5 to thoracic somite 3. The lower limb starts at the level of lumbar somite 2 and finishes between lumbar 5 and sacral 2 (see Figure 24.1). The migrating muscle precursors migrate into the limbs, coalesce and form specific muscle masses which then split to form the definitive muscles of the limbs (see Chapter 24). It is known that, as in skeletal development, cell death is important in the development of these muscle masses. Joints within the limbs develop independently from the musculature (see Chapter 21) but foetal musculature and the motions that occur are required to retain the joint cavities.

Neurons of spinal nerves that follow migrating myoblasts are specific to their original segmental somites. By roughly 9 weeks most muscle groups have formed in their specific locations. The migration of whole myotomes and fusion between them accounts for the grouping of muscular innervation seen in adult limb anatomy.

Movements of the limbs can be detected using ultrasound at 7 weeks and isolated limb movements from around 10–11 weeks.

Head
In the head area the somitomeres undergo similar changes but never fully develop the three compartments of the somite, and this process remains less well understood.

Myogenesis in the head differs from trunk and limb myogenesis as these muscles have different phenotypic properties, although myoblasts still develop from the paraxial mesoderm of the somitomeres and migrate into the pharyngeal arches and their terminal locations.

The surrounding connective tissues coordinate migration and differentiation of muscle as elsewhere, but the nerves to these muscles are present before their formation, as they are cranial nerves. Musculature formed from pharyngeal arches and their innervation is described in Chapters 38–41.

Extraocular muscles probably arise from mesenchyme near the prechordal plate (a thickening of endoderm in the embryonic head). Muscles of the iris are derived from neuroectoderm, whereas ciliary muscle is formed by lateral plate mesoderm. Muscles of the **tongue** form from occipital somites, as does the musculature of the pharynx. Movement of the mouth and tongue and the ability to swallow amniotic fluid begins around week 12.

Smooth muscle
Most smooth muscle of the viscera and gastrointestinal tract (Figure 23.7) is derived from **splanchnic mesoderm** that is located where the organs are developing (Figure 23.8). Developing blood vessels surround local mesenchyme that forms smooth muscle. Larger blood vessels (aorta and pulmonary vessels) receive contributions from **neural crest cells**.

Exceptions to the splanchnic mesoderm rule include muscles of the pupil, erector pili muscles of hair, salivary glands, lacrimal glands, sweat glands and mammary gland smooth muscle, all of which are derived from **ectoderm**.

Cardiac muscle
Cardiac muscle cells are also derived from splanchnic mesoderm surrounding the early heart tube.

The cardiac myoblasts differ from skeletal myoblasts in that they do not fuse to form multinucleated fibres, and they remain individual but connected via intercalated discs (Figure 23.9).

At approximately 22 days a cardiac tube has formed that can contract (see Chapter 25).

Clinical relevance
Muscular dystrophy is a group of over 20 muscular diseases that have genetic causes and all produce progressive weakness and wasting of muscular tissue.

Duchenne muscular dystrophy affects boys (in extremely rare cases symptoms show in female carriers) and affects the gene coding for the protein dystrophin. Patients develop problems with walking between 1 and 3 years of age, wheelchairs are necessary between 8 and 10 years, and life expectancy is limited to late teens to early adulthood as cardiac muscle is affected in the later stages of the disease. There is no cure but research into using stem cells in forms treatment is ongoing.

An absence or partial absence of a skeletal muscle can occur (e.g. **Poland anomaly** which exhibits a unilateral lack of pectoralis major). Other commonly affected muscles include quadriceps femoris, serratus anterior, latissimus dorsi and palmaris longus, and are relatively common.

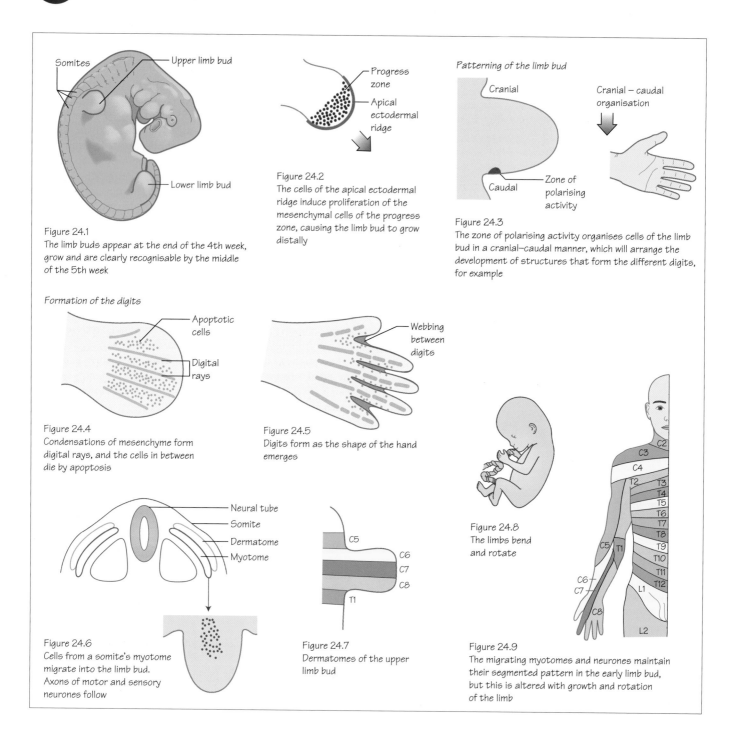

Figure 24.1
The limb buds appear at the end of the 4th week, grow and are clearly recognisable by the middle of the 5th week

Figure 24.2
The cells of the apical ectodermal ridge induce proliferation of the mesenchymal cells of the progress zone, causing the limb bud to grow distally

Figure 24.3
The zone of polarising activity organises cells of the limb bud in a cranial–caudal manner, which will arrange the development of structures that form the different digits, for example

Formation of the digits

Figure 24.4
Condensations of mesenchyme form digital rays, and the cells in between die by apoptosis

Figure 24.5
Digits form as the shape of the hand emerges

Figure 24.6
Cells from a somite's myotome migrate into the limb bud. Axons of motor and sensory neurones follow

Figure 24.7
Dermatomes of the upper limb bud

Figure 24.8
The limbs bend and rotate

Figure 24.9
The migrating myotomes and neurones maintain their segmented pattern in the early limb bud, but this is altered with growth and rotation of the limb

Time period: week 4 to adult

Introduction

Limb development has been studied in great detail, although it is not entirely clear how it is initiated. The mechanisms by which the cells of the early limb are organised, and the fates of those cells,

have been explored for decades, as aberrations of these processes cause gross limb abnormalities.

Limb buds

Cells in the lateral mesoderm at the level of C5–T1 begin to form the upper limb buds at the end of the fourth week and they are

Embryology at a Glance, First Edition. Samuel Webster and Rhiannon de Wreede.

visible from around day 25. The lower limb buds appear a couple of days of later at the level of L1–L5 (Figure 24.1).

Each limb bud has an ectodermal outer covering of epithelium and an inner mesodermal mass of mesenchymal cells.

Distal growth

A series of reciprocal interactions between the underlying mesoderm and overlying ectoderm result in the formation of a thickened ridge of ectoderm called the **apical ectodermal ridge** (AER; Figure 24.2). This ridge forms along the boundary between the dorsal and ventral aspects of the limb bud.

The AER forms on the distal border of the limb and induces proliferation of the underlying cells via fibroblast growth factors (FGF), inducing distal outgrowth of the limb bud. This area of rapidly dividing cells is called the **proliferating zone** (PZ; Figure 24.2). As cells leave the PZ and become further from the AER they begin differentiation and condense into the cartilage precursors of the bones of the limb. Endochondral ossification of these bones is described in Chapter 21.

Organisation

Patterning within the early limb bud controls the proliferation and differentiation of mesenchymal cells, forming the structures of the limb. The AER controls the proximal–distal axis, for example.

A group of cells in the caudal mesenchyme of the limb bud act as a **zone of polarising activity** (ZPA; Figure 24.3), secreting a morphogen that diffuses cranially and themselves contributing to development of the digits. The ZPA has a role in a cranial–caudal axis (i.e. specifying where the thumb and little finger form; Figure 24.3).

The dorsal–ventral axis is controlled by signals from the dorsal and ventral ectoderm. These signals specify which side of the hand the nails should form on and which side the fingertips, for example.

Disruption of these patterning signals (and others) causes limb malformations.

Digits

During weeks 6 and 7 (development of the lower limbs lags behind that of the upper limbs) the distal edges of the limb buds flatten to form **hand** and **foot plates**. Digits begin to develop as condensations of mesenchymal cells clump together to construct long thickenings (Figure 24.4). Localised programmed cell death between these digit primordia splits the plate into five digital rays, and the mesenchymal condensations develop to become the bones and joints of the phalanges (Figures 24.4 and 24.5).

Dermatomes and myotomes

Cells from the dermamyotomes of somites (see Chapter 20) at the levels of the limb buds migrate into the limbs, and differentiate into myoblasts. They group to form dorsal and ventral masses, which will approximate to the muscles of the flexor and extensor compartments of the adult.

Motor neurons from the ventral rami of the spinal cord at the levels of the limb buds (C5–T1 for the upper limbs, L4–S3 for the lower limbs) extend axons into the limbs, following the myoblasts (Figure 24.6). Control of this axon growth also occurs independent of muscle development, however. Dorsal branches from each ventral ramus pass to muscles of the dorsal mass (extensors), and ventral branches from each ventral ramus pass to the ventral mass (flexors). Also, more cranial neurons (C5–C7 in the upper limb, for example) pass to craniodorsal parts of the limb bud, and more caudal neurons (C8–T2) pass to ventrocaudal parts.

As axons enter the limb bud they mix to create the brachial and lumbosacral plexuses during this development stage, before the axons continue onwards to their target muscles. Branches combine to form larger dorsal and ventral nerves, eventually the radial, musculocutaneous, ulnar and median nerves in the upper limb, for example. The radial nerve forms from dorsal branches, as it is a nerve that innervates the extensor muscles of the upper arm and forearm.

The muscle groups, initially neatly organised, fuse and adult muscles may be derived from myoblasts from multiple somites. Likewise, axons of the dorsal root ganglia initially carry sensory innervation from the skin of the limb in an organised pattern of dermatomes.

The upper limb begins to become flexed at the elbow, and the lower limb develops a bend at the knee in week 7. The limbs also rotate, transforming from a simple, outwardly extending limb bud to a more recognisable limb shape. The upper limb rotates laterally by 90° and the lower limb rotates medially by 90° (Figure 24.7). By the end of week 8 the upper and lower limbs are well defined, with pads on the fingers and toes. The hands meet in the midline, and the feet have become close together.

With the rotation and bending of the limbs, and the fusing of early muscles, the patterns of muscle innervation and dermatomes are disrupted and produce the adult patterns (Figures 24.7–24.9).

Clinical relevance

The period of early limb development of weeks 4 and 5 is susceptible to interruption by teratogens, as seen in the thalidomide epidemic of congenital limb abnormalities of the 1950s and 1960s. The earlier the teratogen is applied to the foetus, the more severe the developmental defects.

Achondroplastic dwarfism is caused by a mutation in the fibroblast growth factor receptor 3 gene (*FGFR3*). FGF signalling via this receptor is involved in growth plate function, and disruption of this causes limited long bone growth and disproportionate short stature.

Meromelia describes the partial absence of a limb, and **amelia** the complete absence of a limb. **Phocomelia** refers to a limb in which the proximal part is shortened, and the hand or foot is attached to the torso by a shortened limb.

In **polydactyly** an extra digit, often incomplete, forms on the hand or foot. **Ectrodactyly** describes missing digits, and often lateral digits forming a claw-shaped hand or foot. A hand or foot with **brachydactyly** has shortened digits. A person with **syndactyly** has webbed digits as the interdigital cells failed to apoptose normally.

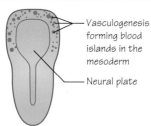

Vasculogenesis forming blood islands in the mesoderm

Neural plate

Figure 25.1
Blood islands appear in the lateral plate mesoderm from angioblasts that join together as a syncytium (week 3)

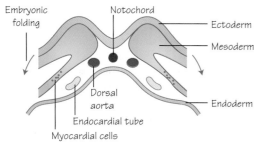

Embryonic folding

Notochord

Ectoderm

Mesoderm

Dorsal aorta

Endocardial tube

Myocardial cells

Endoderm

Figure 25.2
Location of the endocardial tube and myocardial cells in the embryo before the embryo begins folding. Transverse section

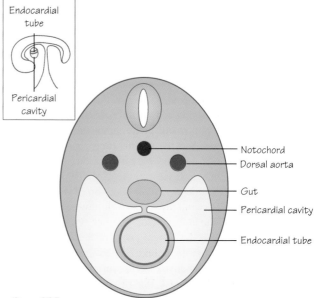

Endocardial tube

Pericardial cavity

Notochord

Dorsal aorta

Gut

Pericardial cavity

Endocardial tube

Figure 25.3
Anterior position of the endocardial tube surrounded by the pericardial cavity relative to the gut, in cross section at 22 days.
Insert: Region of cross section

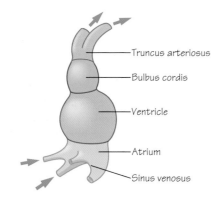

Truncus arteriosus

Bulbus cordis

Ventricle

Atrium

Sinus venosus

Figure 25.4
The early heart tube (22 days)

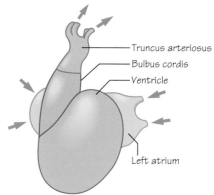

Truncus arteriosus

Bulbus cordis

Ventricle

Left atrium

Figure 25.5
The folded heart tube (29 days)

Embryology at a Glance, First Edition. Samuel Webster and Rhiannon de Wreede.

Time period: days 16–28

Formation of the heart tube

During the third week of development blood islands appear in the lateral plate mesoderm (Figure 25.1) from angioblasts that accumulate as a **syncytium** (rather like the formation of the syncytiotrophoblast that we saw form during the development of the placenta in Chapter 12). From these cells new blood cells and blood vessels form through vasculogenesis. Blood islands at the cranial end of the embryo merge and assemble a horseshoe-shaped tube lined with endothelial cells which curves around the embryo in the plane of the mesoderm.

Progenitor cells that migrated from the epiblast differentiate in response to signals from the nearby endoderm to become myoblasts and surround the horseshoe-shaped tube (Figure 25.2). This developing cardiovascular tissue is called the **cardiogenic field**.

The early heart tube expands into the newly forming pericardial cavity (Figure 25.3) as it begins to link with the paired dorsal aortae cranially and veins caudally. The developing central nervous system and folding of the embryo (see Chapter 18) pushes it into the thorax and brings the developing parts of the cardiovascular system towards one another (Figures 25.1–25.3).

Looping and folding of the heart tube

The early, simple heart tube (Figure 25.4) undergoes a series of foldings to bring it from a straight tube to a folded shape ready to become four chambers. The heart tube begins to bend at 23 days (stops at 28 days) and develops two bulges. The cranial bulge is called the **bulbus cordis** and the caudal one is the **primitive ventricle** (Figure 25.5). These continue to bend and create the **cardiac** (or bulboventricular) **loop** during the fourth week of development.

When the heart tube loops, the top bends towards the right so that the bulboventricular part of the heart becomes U-shaped. This looping changes the anterior–posterior polarity of the heart into the left–right that we see in the adult. The bulbus cordis forms the right part of the 'U' and the primitive ventricle the left part. You can see the junction between the bulbus cordis and ventricle by the presence of the **bulboventricular sulcus**. The looping causes the atrium and sinus venosus to move dorsal to the heart loop.

The atrium is now dorsal to the other parts of the heart and the common atrium is connected to the primitive ventricle by the **atrioventricular canal**. The primitive ventricle will develop into most of the left ventricle and the proximal section of the bulbus cordis will form much of the right ventricle. The **conus cordis** will form parts of the ventricles and their outflow tracts, and the **truncus arteriosus** will form the roots of both great vessels.

Sinus venosus (right atrium)

The **sinus venosus** comprises the inflow to the primitive heart tube and is formed by the major embryonic veins (common cardinal, umbilical and vitelline) as they converge at the right and left sinus horns (see Chapter 28). The sinus venosus passes blood from the veins to the primitive atrium.

With time, venous drainage becomes prioritised to the right side of the embryo and the left sinus horn becomes smaller and less significant, eventually forming the coronary sinus and draining the coronary veins into the right atrium. The right sinus horn persists, enlarges and becomes part of the **inferior vena cava** entering the heart and incorporated into the right atrium, forming much of its wall.

Similarly, a single pulmonary vein is initially connected to the left side of the primitive atrium and divides twice during the fourth week to form four pulmonary veins. These become incorporated into the wall of the future left atrium and extend towards the developing lungs.

Clinical relevance

Many congenital heart defects occur later in development during the division of the heart into its four chambers.

Dextrocardia is a condition in which the heart lies on the right, with the apex of the left ventricle pointing to the right, instead of the left. This is often associated with **situs inversus**, a condition in which all organs are asymmetrical. Other congenital heart defects can occur with dextrocardia but it is often asymptomatic.

Circulatory system: heart chambers

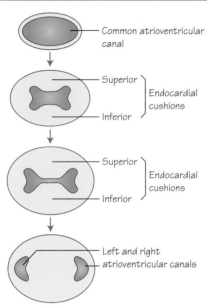

Common atrioventricular canal

Superior ⎱ Endocardial
Inferior ⎰ cushions

Superior ⎱ Endocardial
Inferior ⎰ cushions

Left and right atrioventricular canals

Figure 26.1
The endocardial cushions split the single atrioventricular canal into 2 canals linking the atrium and ventricle (weeks 5 and 6)

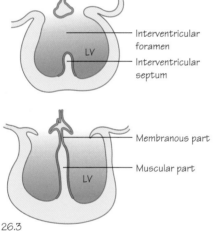

Interventricular foramen

Interventricular septum

Membranous part

Muscular part

Figure 26.3
The formation of the interventricular septum (weeks 5 to 7)

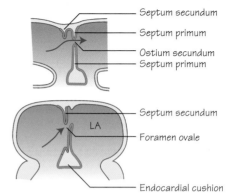

Septum secundum
Septum primum
Ostium secundum
Septum primum

Septum secundum
Foramen ovale

Endocardial cushion

Figure 26.2
The formation of the atrial septa (weeks 5 and 6)

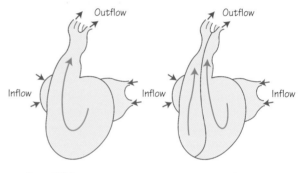

Outflow

Outflow

Inflow

Inflow

Inflow

Figure 26.4
The single outflow tract of the conus arteriosus and truncus arteriosus is split into 2 by the conotruncal septum

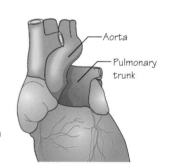

Aorta

Pulmonary trunk

Figure 26.5
The adult pulmonary trunk and aorta twist around each other as they rise superiorly from the ventricles

Embryology at a Glance, First Edition. Samuel Webster and Rhiannon de Wreede.

Time period: day 22

Dividing the heart into chambers

Heart septa appear during week 5 and divide the heart tube into four chambers between days 27 and 37. The septa form as inward growths of endocardium separating the atrial and ventricular chambers, splitting the atrium into left and right, and splitting the ventricle and bulbus cordis into left and right ventricles, respectively (Figure 26.1).

The **atrioventricular canal** connects the primitive atrium and ventricle. At the end of week 4 the endocardium of the anterior and posterior walls of the atrioventricular canal thicken and bulge outwards into the canal's lumen. These are the **endocardial cushions** and by the end of week 6 they meet in the middle, splitting the atrioventricular canal into two canals (Figure 26.1).

Atria

At the same time, new tissue forms in the roof of the primitive atrium. This thin, curved septum is the **septum primum** and extends down from the roof, growing towards the endocardial cushions. The primitive atrium begins to split into left and right atria. The gap remaining inferior to the septum primum is the **ostium primum** (Figure 26.2). Growth of the endocardial cushions and the septum primum cause them to meet.

A second ridge of tissue grows from the roof of the atrium, on the right side of the septum primum. This is called the **septum secundum** (Figure 26.2) and grows towards the endocardial cushions, but stops short. The gap remaining is the **ostium secundum**, and the two holes and flap of the septum primum against septum secundum form a one-way valve allowing blood to shunt from the right atrium to the left but not in reverse. This is the **foramen ovale** (Figure 26.2) and is one of the routes that exist before birth allowing blood circulation to circumvent the developing lungs. A change in pressure between atria at birth holds the septum primum closed against the septum secundum, and the foramen becomes permanently sealed.

Ventricles

From the end of the fourth week a **muscular interventricular septum** arises from the floor of the ventricular chamber as the two primitive ventricles begin to expand (Figure 26.3). The septum rises towards the endocardial cushions, leaving an **interventricular foramen**. As the atrioventricular septum is completed late in the seventh week the endocardial cushion extends inferiorly (as the **membranous interventricular septum**) to complete the **interventricular septum** and close the interventricular foramen (Figure 26.3).

Now the heart is four connected chambers with two input tubes. The single outflow tract of the primitive heart must also split into two to pass blood from the ventricles to the pulmonary and systemic circulatory systems (Figure 26.4). The **conotruncal** outflow tract, comprising the **conus arteriosus** and **truncus arteriosus**, develops a pair of longitudinal ridges on its internal surface. These grow towards one another and fuse to form the **conotruncal septum**, which meets with the muscular interventricular septum to link each ventricle with its outflow artery. The conotruncal septum spirals within the conus arteriosus and truncus arteriosus, giving the intertwining nature of the adult pulmonary trunk and aorta (Figure 26.5).

Valves

After the fusion of the endocardial cushions to form two atrioventricular canals, mesenchymal cells proliferate in the walls of the canals. The ventricular walls inferior to this erode, leaving leaflets of primitive valves and thin connections to the walls of the ventricles. These connections develop into the fibrous **chordae tendinae** with papillary muscles at their ventricular ends. The **left atrioventricular valve** develops two leaflets (the bicuspid valve) and the **right atrioventricular valve** usually develops three (the tricuspid valve).

The semilunar valves of the aorta and pulmonary trunk develop in a similar manner during the formation of the **conotruncal septum**.

Neural crest cells

Neural crest cells, appearing during neurulation, migrate from the developing neural tube to take part in the development of an astounding range of different structures, including the heart. In the heart they contribute to the conotruncal septum.

Clinical relevance

Heart defects are the most common congenital defects, generally occurring because of problems with structural development processes. Six in 1000 children are born with a heart defect.

A **ventricular septal** defect is the most common heart defect, and failure of the membranous interventricular septum to close completely allows blood to pass from the left to right ventricles. Most will close on their own but surgery may be required. This can be linked to other conotruncal defects. **Atrial septal** defects occur when the foramen ovale fails to close (patent foramen ovale), allowing blood to pass between atria after birth. Treatment is surgical.

Abnormal narrowing of the pulmonary or aortic valves can give pulmonary or **aortic stenosis**, forcing the heart to work harder. Stenosis of the aorta will limit the systemic circulation, with clear consequences. These arteries can be transposed if the conotruncal septum fails to form its spiral course, and the aorta will arise from the right ventricle and the pulmonary trunk from the left ventricle (transposition of the great vessels). Low oxygen blood is passed into the systemic circulation.

Tetralogy of Fallot describes four congenital defects resulting from abnormal development of the conotruncal septum: pulmonary stenosis, an overriding aorta connected to both ventricles, a ventricular septal defect and hypertrophy of the wall of the right ventricle. Poorly oxygenated blood is pumped in the systemic circulation with symptoms of cyanosis and breathlessness. Surgical intervention is required.

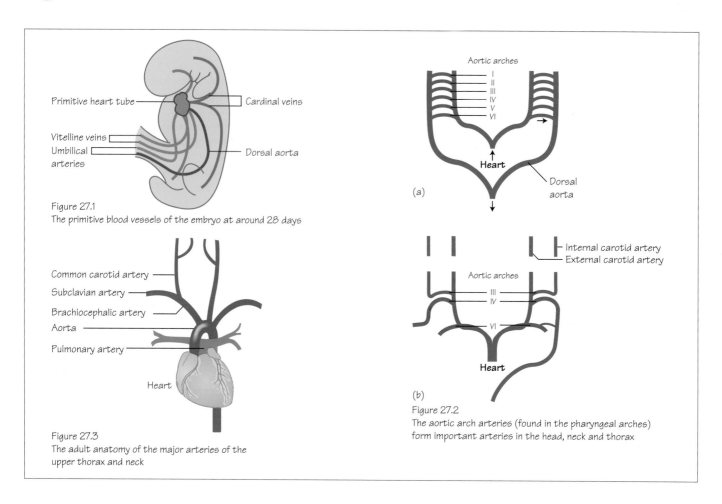

Figure 27.1
The primitive blood vessels of the embryo at around 28 days

Primitive heart tube
Cardinal veins
Vitelline veins
Umbilical arteries
Dorsal aorta

(a)

Aortic arches
I
II
III
IV
V
VI
Heart
Dorsal aorta

Common carotid artery
Subclavian artery
Brachiocephalic artery
Aorta
Pulmonary artery
Heart

Figure 27.3
The adult anatomy of the major arteries of the upper thorax and neck

Internal carotid artery
External carotid artery
Aortic arches
III
IV
VI
Heart

(b)
Figure 27.2
The aortic arch arteries (found in the pharyngeal arches) form important arteries in the head, neck and thorax

Embryology at a Glance, First Edition. Samuel Webster and Rhiannon de Wreede.

Time period: day 18 to birth

Vasculogenesis

Vasculogenesis is the formation of new blood vessels from cells that were not blood vessels before. As if by magic, blood cells and vessels appear in the early embryo. In fact, mesodermal cells are induced to differentiate into haemangioblasts, which further differentiate into both haematopoietic stem cells and angioblasts. Haematopoietic stem cells will form all the blood cell types, and angioblasts will build the blood vessels. Separate sites of vasculogenesis may merge to form a network of blood vessels, or new vessels may grow from existing vessels by angiogenesis. When the liver forms it will be the primary source of new haematopoietic stem cells during development.

Angiogenesis

Angiogenesis is the development of new blood vessels from existing vessels. Endothelial cells detach and proliferate to form new capillaries. This process is under the influence of various chemical and mechanical factors. Although important in growth this also occurs in wound healing and tumour growth, and as such angiogenesis has become a target for anti-cancer drugs.

Primitive circulation

Near the end of the third week blood islands form through vasculogenesis on either side of the cardiogenic field and the notochord (see Chapter 25). They merge, creating two lateral vessels called the **dorsal aortae** (Figure 27.1). These blood vessels receive blood from three pairs of veins, including the **vitelline veins** of the yolk sac (a site of blood vessel formation external to the embryo), the **cardinal veins** and the **umbilical veins** (Figure 27.1).

Blood flows from the dorsal aortae into the **umbilical arteries** and the **vitelline arteries**. Branches of the dorsal aortae later fuse to become the single descending aorta in adult life.

The heart tube will form where veins drain to the dorsal aortae. The **aortic arches** within the pharyngeal arches form here, linking the outflow of the primitive heart to the dorsal aortae. Blood flow begins during the fourth week.

Aortic arches

Five pairs of aortic arches form between the most distal part of the truncus arteriosus and the dorsal aortae. They develop within the pharyngeal arches during weeks 4 and 5 of development and are associated with other structures derived from the pharyngeal arches in the head and neck.

The aortic arches grow in sequence and therefore are not all present at the same time. One little mystery in embryology is that the fifth aortic arch (and pharyngeal arch) either does not form or it grows and then regresses. For that reason the five aortic arch arteries that do develop are named I, II, III, IV and VI (Figure 27.2).

The **truncus arteriosus** also divides and develops into the ventral part of the aorta and pulmonary trunk. Its most distal part forms left and right horns that also contribute to the brachiocephalic artery.

The five aortic arches and paired dorsal aortae combine and develop into a number of vessels of the head and neck (Figure 27.3):

Aortic arch I Maxillary artery
Aortic arch II Stapedial artery (rare)
Aortic arch III Common carotid artery and internal carotid artery (external carotid artery is an angiogenic branch of aortic arch III)
Aortic arch IV Right side, right subclavian artery (proximal portion)
Left side, aortic arch (portion between the left common carotid and subclavian arteries)
Aortic arch VI Right side, right pulmonary artery
Left side, left pulmonary artery and ductus arteriosus

Ductus arteriosus

Aortic arch VI forms as a link between the truncus arteriosus and the left dorsal aorta (Figure 27.2); this link persists until birth as the ductus arteriosus. This vessel allows blood flow to bypass the lungs as it connects the pulmonary trunk with the aorta. Foetal pulmonary vascular resistance is high and most blood from the right ventricle (85–90%) passes through the ductus arteriosus to the aorta. Blood flow to the lungs is minimal during gestation and they are protected from circulatory pressures during development. This shunt also allows the wall of the left ventricle to thicken.

Coronary arteries

The blood supply to the tissue of the heart has been considered to form by angiogenesis from the walls of the right and left aortic sinuses (bulges in the aorta that occur just superior to the aortic valve). This may be influenced by specific tension in the walls of the heart. Vessels form that link with a plexus of epicardial vessels on the surface of the heart. The reverse may be true, however, and these arteries may grow from the epicardial plexus into the aorta and right atrium to initiate their function. Recently, cells from the sinus venosus have been tracked as angiogenic sprouts that migrate over the myocardium and form both coronary arteries and veins and these cells may, in fact, be the source of all the coronary blood vessels.

Clinical relevance

Coarctation of the aorta is a narrowing of the aorta sometimes found distal to the point from which the left subclavian artery arises. It may be described as preductal or postductal depending upon its location relative to the ductus arteriosus. With postductal coarctation, a collateral circulation develops linking the aorta proximal to the ductus arteriosus with inferior arteries. With a preductal coarctation the route of blood flow through the ductus arteriosus to inferior parts of the body is lost with birth causing hypoperfusion of the lower body.

Aberrations in aortic arch development may give anomalous arteries, such as a right arch of the aorta or a vascular ring around the trachea and oesophagus.

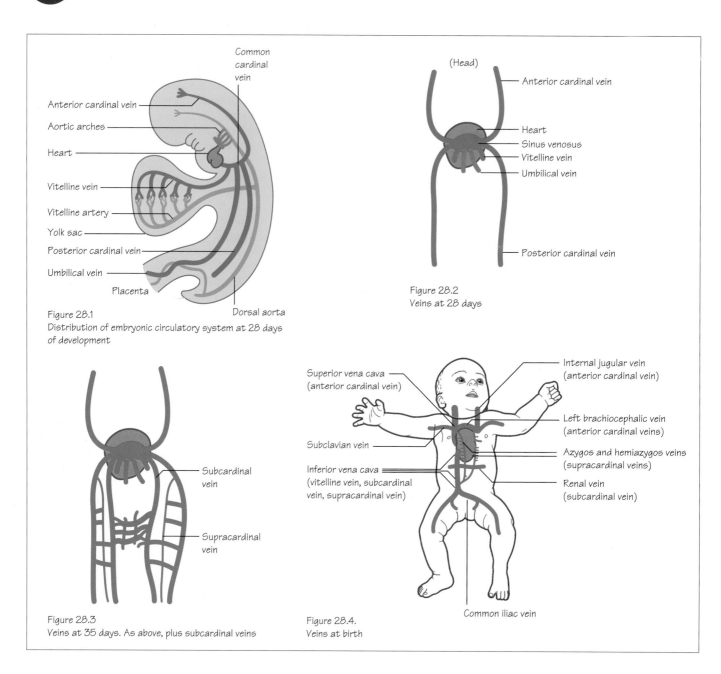

Figure 28.1
Distribution of embryonic circulatory system at 28 days of development

Common cardinal vein
Anterior cardinal vein
Aortic arches
Heart
Vitelline vein
Vitelline artery
Yolk sac
Posterior cardinal vein
Umbilical vein
Placenta
Dorsal aorta

Figure 28.2
Veins at 28 days

(Head)
Anterior cardinal vein
Heart
Sinus venosus
Vitelline vein
Umbilical vein
Posterior cardinal vein

Figure 28.3
Veins at 35 days. As above, plus subcardinal veins

Subcardinal vein
Supracardinal vein

Figure 28.4.
Veins at birth

Superior vena cava (anterior cardinal vein)
Subclavian vein
Inferior vena cava (vitelline vein, subcardinal vein, supracardinal vein)
Internal jugular vein (anterior cardinal vein)
Left brachiocephalic vein (anterior cardinal veins)
Azygos and hemiazygos veins (supracardinal veins)
Renal vein (subcardinal vein)
Common iliac vein

Embryology at a Glance, First Edition. Samuel Webster and Rhiannon de Wreede.

64 © 2012 John Wiley & Sons, Ltd. Published 2012 by John Wiley & Sons, Ltd.

Time period: day 18 to birth

Vitelline vessels

The vitelline circulation is the flow of blood between the embryo and the yolk sac through a collection of vitelline arteries and veins that pass within the yolk stalk (Figure 28.1).

The vitelline arteries are branches of the dorsal aortae, and most of them degenerate in time. Those that remain fuse and form the 3 unpaired ventral arterial branches of the aorta that supply the gut: the celiac trunk, superior mesenteric artery and inferior mesenteric artery.

The vitelline veins will give rise to the hepatic portal vein and the hepatic veins of the liver.

Umbilical vessels

The umbilical circulation is the flow of blood between the chorion of the placenta and the embryo. The umbilical arteries carry poorly oxygenated blood to the placenta and the veins carry highly oxygenated blood initially to the heart of the embryo (Figure 28.1), and later into the liver when it forms (see Figure 29.1). The right umbilical vein is lost around week 7, leaving only the left to carry blood from the placenta.

The formation of the **ductus venosus** during the foetal period causes about half of the blood from the umbilical vein to flow directly into the inferior vena cava, bypassing the liver (Figure 29.1). This, with other mechanisms, preferentially shunts highly oxygenated blood to the foetal brain.

Of the umbilical arteries only the proximal portions persist as parts of the internal iliac arteries and superior vesical arteries in the adult. The distal portions do not remain as arteries but become the medial umbilical ligaments. The umbilical vein becomes the ligamentum teres, passing from the umbilicus to the porta hepatis in the adult (see Chapter 29).

Cardinal veins

The common cardinal veins initially form an H-shaped structure, with the horizontal bar being the sinus venosus that links the major veins and the atrium of the early heart tube (Figure 28.2). The left and right anterior (or superior) branches drain blood from the head and shoulder regions and the posterior (or inferior) branches drain from the abdomen, pelvis and lower limbs.

At 6 weeks a **subcardinal** vein arises on either side of the embryo caudal to the heart and anastomoses with the posterior cardinal veins (Figure 28.3). The subcardinal veins also form an anastomosis with each other anterior to the dorsal aortae, and tributaries are sent into the developing limbs. The right subcardinal vein joins vessels of the liver. Similarly, at 7 weeks **supracardinal** veins form and link to the posterior cardinal veins (Figure 28.3).

The posterior cardinal veins degenerate, although the most caudal parts continue as a sacral venous plexus and later as the common iliac veins.

An important junction between the right supracardinal and right subcardinal vein forms and both will become sections of the inferior vena cava (IVC). Parts of the right posterior cardinal veins, common, subcardinal and supracardinal veins also contribute. A shift towards the right side occurs, with degeneration of venous structures on the left side and the formation and enlargement of the inferior vena cava on the right (Figure 28.4).

Similarly, the degeneration of much of the left anterior cardinal vein gives a shift to the right side as the right anterior cardinal vein forms part of the superior vena cava (SVC) and the right brachiocephalic vein (Figure 28.4). An anastomosis between the 2 anterior cardinal veins persists as the left brachiocephalic vein.

The right supracardinal vein becomes much of the **azygos vein**, and the left supracardinal vein forms part of the **hemiazygos vein** and the accessory hemiazygos veins (Figure 28.4). Branches from the subcardinal vein network form renal, suprarenal and the gonadal veins.

Clinical relevance

The formation of the venous system is somewhat variable and complicated, and can give rise to variations in adult SVC and IVC anatomy. The hepatic section of the IVC may fail to form, for example, and blood instead flows back to the heart through the azygos and hemiazygos veins from the inferior parts of the body (**azygos continuation**). Persistence of supracardinal veins can leave **double inferior vena cavae**, and persistence of the left anterior cardinal vein can give **double SVC**. In this case the right anterior vena cava may even degenerate, leaving only a **left SVC.** These variations are not common.

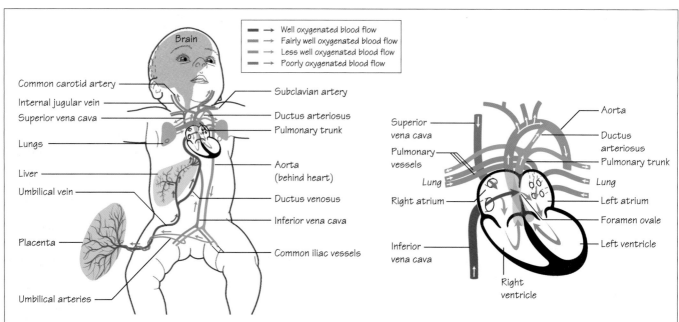

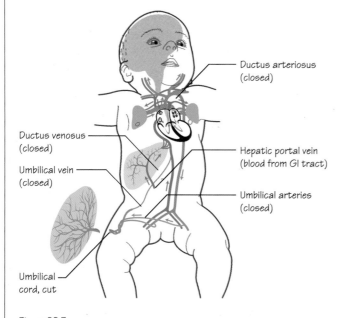

Figure 29.1
The foetal circulatory system. Half of the blood from the umbilical vein bypasses the liver via the ductus venosus. Oxygen saturation of the blood leaving the heart is reduced by blood entering from the superior vena cava and the coronary sinus

Figure 29.2
The foetal circulation, a closer view of the heart

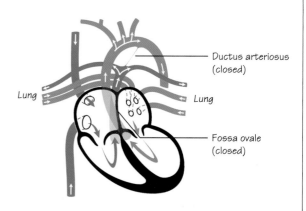

Figure 29.5
Neonatal circulation. At birth the lungs begin to function, the ductus arteriosus and ductus venosus close, and the umbilical vessels close

Figure 29.4
Neonatal circulation, a closer view of the heart

Embryology at a Glance, First Edition. Samuel Webster and Rhiannon de Wreede.

Time period: birth (38 weeks)

Foetal blood circulation

Dramatic and clinically significant changes occur to the circulatory and respiratory systems at birth. Here, we look at changes primarily of the circulatory system and how these changes prepare the baby for life outside the uterus.

If we were to follow the flow of oxygenated blood in the foetus from the placenta (Figure 29.1), we would start in the umbilical vein and track the blood moving towards the liver. Here, half the blood enters the liver itself and half is redirected by the **ductus venosus** directly into the inferior vena cava, bypassing the liver. The blood remains well oxygenated and continues to the right atrium, from which it may pass into the right ventricle in the expected manner or directly into the left atrium via the **foramen ovale** (Figure 29.2). Blood within the left atrium passes to the left ventricle and then into the aorta.

Blood entering the right atrium from the superior vena cava and the coronary sinus is relatively poorly oxygenated. The small amount of blood that returns from the lungs to the left atrium is also poorly oxygenated. Mixing of this blood with the well-oxygenated blood from the ductus venosus reduces the oxygen saturation somewhat.

Blood within the right ventricle will leave the heart within the pulmonary artery, but most of that blood will pass through the **ductus arteriosus** and into the descending aorta. Almost all of the well-oxygenated blood that entered the right side of the heart has avoided entering the pulmonary circulation of the lungs, and has instead passed to the developing brain and other parts of the body (Figure 29.3).

Ductus venosus

The umbilical arteries constrict after birth, preventing blood loss from the neonate. The umbilical cord is not cut and clipped immediately after birth, however, allowing blood to pass from the placenta back to the neonatal circulation through the umbilical vein.

The ductus venosus shunted blood from the umbilical vein to the inferior vena cava during foetal life, bypassing the liver. After birth a sphincter at the umbilical vein end of the ductus venosus closes (Figure 29.4). The ductus venosus will slowly degenerate and become the **ligamentum venosus**.

Once the umbilical circulation is terminated the umbilical vein will also degenerate and become the **round ligament** (or ligamentum teres hepatis) of the liver. This may be continuous with the ligamentum venosus. The umbilical arteries will persist in part as the superior vesical arteries, supplying the bladder, and the remainder will degenerate and become the **median umbilical ligaments**.

Ductus arteriosus

The shunt formed by the ductus arteriosus between the pulmonary trunk and the aorta in foetal life causes blood rich in oxygen to bypass the lungs, which have a very high vascular resistance during development. With birth, the first breath of air and early use of the lungs the pulmonary vascular resistance drops and blood flow to the lungs increases. An increase in oxygen saturation of the blood, bradykinin produced by the lungs, and a reduction in circulating prostaglandins cause the smooth muscle of the wall of the ductus arteriosus to contract, restricting blood flow here and increasing blood flow through the pulmonary arteries (Figure 29.4). Physiological closure is normally achieved within 15 hours of birth.

During the first few months of life, the ductus arteriosus closes anatomically, leaving the **ligamentum arteriosum** as a remnant. As this is a remnant of the sixth aortic arch the left recurrent laryngeal nerve can be found here (see Chapter 41).

Foramen ovale

The direction in which blood flows into the right atrium from the inferior vena cava and the crista dividens (the lower edge of the septum secundum, forming the superior edge of the foramen ovale) preferentially direct the flow of blood through the foramen ovale into the left atrium, reducing mixing with poorly oxygenated blood entering the right atrium from the superior vena cava (Figures 29.2 and 29.3).

As the child takes his or her first breath the reduction in pulmonary vascular resistance and subsequent flow of blood through the pulmonary circulation increases the pressure in the left atrium. As the pressure in the left atrium is now higher than in the right atrium the septum primum is pushed up again the septum secundum, thus functionally closing the foramen ovale (Figure 26.3). Anatomical closure is usually completed within the next 6 months. In the adult heart a depression called the **fossa ovalis** remains upon the interior of the right atrium.

Clinical relevance

Patent foramen ovale (PFO) is an atrial septal defect. The foramen ovale fails to close anatomically although it is held closed by the difference in interatrial pressure. A 'backflow' of blood can occur from left to right under certain circumstances which increases pressure in the thorax. These circumstances include sneezing or coughing, and even straining during a bowel movement. Autopsy studies have shown a PFO incidence of 27% in the US population but those with this defect generally do not have symptoms. Treatment varies depending upon age and associated problems, but often no treatment is necessary.

If the ductus arteriosus fails to close at birth it is termed a **patent ductus arteriosus** (PDA). Well-oxygenated blood from the aorta mixes with poorly oxygenated blood from the pulmonary arteries, causing tachypnoea, tachycardia, cyanosis, a widened pulse pressure and other symptoms. Longer term symptoms seen during the first year of life include poor weight gain and continued laboured breathing. Premature infants are more likely to develop a PDA. Treatment can be surgical or pharmacological.

A **portosystemic shunt** is less common and occurs when the ductus venosus fails to close at birth, allowing blood to continue to bypass the liver. A build-up of uric acid and ammonia in the blood can lead to a failure to gain weight, vomiting and impaired brain function.

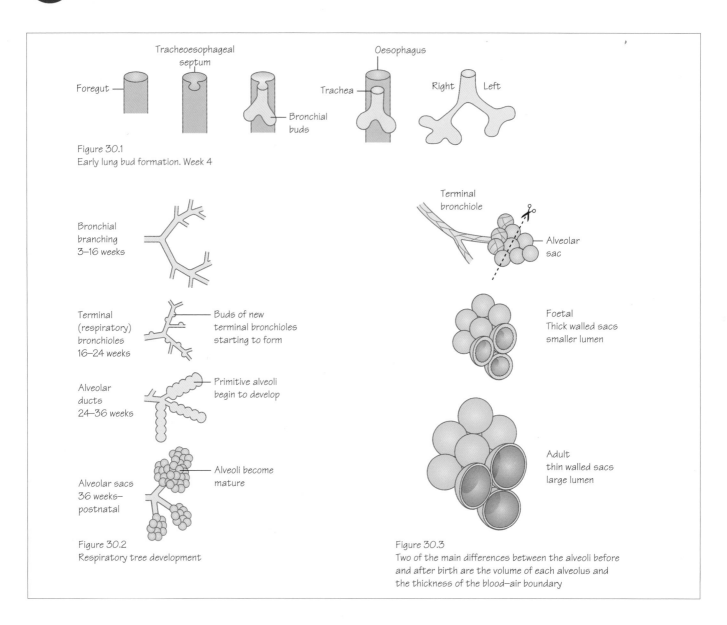

Figure 30.1
Early lung bud formation. Week 4

Figure 30.2
Respiratory tree development

Figure 30.3
Two of the main differences between the alveoli before and after birth are the volume of each alveolus and the thickness of the blood–air boundary

Time period: day 28 to childhood

Introduction

The development of the respiratory system is continuous from the fourth week, when the respiratory diverticulum appears, to term. The 24-week potential viability of a foetus (approximately 50% chance of survival) is partly because at this stage the lungs have developed enough to oxygenate the blood. Limiters to oxygenation include the surface area available to gaseous exchange, the vascularisation of those tissues of gaseous exchange and the action of surfactant in reducing the surface tension of fluids within the lungs.

Development of the respiratory system includes not only the lungs, but also the conducting pathways, including the trachea, bronchi and bronchioles. Lung development can be described in

five stages: **embryonic, pseudoglandular, canalicular, saccular** and **alveolar**.

Although not in use as gas exchange organs *in utero*, the lungs have a role in the production of some amniotic fluid.

Lung bud

The development of the respiratory system begins with the growth of an endodermal bud from the ventral wall of the developing gut tube in the fourth week (Figure 30.1).

To separate the lung bud from the gut tube two longitudinal folds form in the early tube of the foregut, meet and fuse, creating the **tracheoesophageal septum**. This division splits the dorsal foregut (oesophagus) from the ventral lung bud (larynx, trachea and lung). These structures remain in communication superiorly through the laryngeal orifice.

Embryology at a Glance, First Edition. Samuel Webster and Rhiannon de Wreede.

Being derived from the gut the epithelial lining is endodermal in origin, but as the bud grows into the surrounding mesoderm reciprocal interactions between the germ layers occur. The mesoderm develops into the cartilage and smooth muscle of the respiratory conduction pathways.

Respiratory tree

In the fifth week the tracheal bud splits and forms two lateral outgrowths: the **bronchial buds**. It is at this early stage we see the asymmetry of the lungs appear; the right bud forms three bronchi and the left two. The bronchial buds branch and extend, forming the respiratory tree of the three right lobes and two left lobes of the lungs (Figure 30.1).

Up to week 5 the first period of lung development is known as the **embryonic stage**.

From 6 weeks their development enters the **pseudoglandular stage**. The respiratory tree continues to lengthen and divide with 16–20 generations of divisions by the end of this stage (Figure 30.2). Histologically, the lungs resemble a gland at this stage.

Epithelial cells of the bronchial tree become ciliated and the beginnings of respiratory elements appear. Cartilage and smooth muscle cells appear in the walls of the bronchi. Lung-specific **type II alveolar cells** (pneumocytes) begin to appear. These are the cells that will produce surfactant.

The pseudoglandular stage ends at approximately 16 weeks, by which time the entire respiratory tree, including terminal bronchioles, has formed (Figure 30.2).

Alveoli

During the next phase, known as the **canalicular stage** (17–24 weeks), the respiratory parts of the lungs develop. Canaliculi (canals or tubes) branch out from the terminal bronchioles. Each forms an **acinus** comprising the terminal bronchiole, an alveolar duct and a terminal sac (Figure 30.2). This is the primitive **alveolus**.

The duct lumens become wider and the epithelial cells of some of the primitive alveoli flatten to form **type I alveolar cells** (also known as type I pneumocytes, or squamous alveolar cells). These will be the cells of gaseous exchange.

An invasion of capillaries into the mesenchyme surrounding the primitive alveoli brings blood vessels to the type I alveolar cells. Towards the end of the canalicular stage some primitive alveoli are sufficiently developed and vascularised to allow gaseous exchange, and a foetus born at this stage may survive with intensive care support.

The **saccular stage** (or terminal sac period, from 25 weeks to birth), describes the continued development of the respiratory parts of the lungs. **Type II alveolar cells** (also known as type II pneumocytes, great alveolar cells or septal cells) begin to produce **surfactant**, a phospholipoprotein that reduces the surface tension of the fluid in the lungs and will prevent collapse of the alveoli upon expiration and improve lung compliance after birth.

During this stage many more primitive alveolar sacs develop from the terminal bronchioles and alveolar ducts. The **blood–air barrier** between the epithelial type I alveolar cells and endothelial cells of the capillaries develops in earnest, and the surface area available to gaseous exchange begins to increase considerably.

Table 30.1 Stages in the development of the respiratory system

Stage	Time	Development
Embryonic	3–5 weeks	Initial bud and branching
Pseudoglandular	6–16 weeks	Complete branching
Canalicular	17–24 weeks	Terminal bronchioles
Saccular	25 weeks to term	Terminal sacs and capillaries cone into close contact
Alveolar	8 months to childhood	Well-developed blood–air barrier

The final **alveolar stage** (36 weeks onwards) begins a few weeks before birth and continues postnatally through childhood. Alveoli increase in number and diameter enlarging the surface area available to gas exchange (Figure 30.2). The squamous (type I alveolar) epithelial cells lining the primitive alveoli continue to thin before birth, forming **mature alveoli** (Figure 30.3). Septation divides the alveoli. Surfactant is produced in sufficient quantities for normal lung function with birth. Continued development through childhood will increase the number of alveoli from 20–50 million at birth to around 400 million in the adult lung (Table 30.1).

Circulation

Two classes of blood circulation are present in the lungs: pulmonary and bronchial. Pulmonary arteries derive from the artery of the sixth pharyngeal arch and accompany the bronchial tree as it branches, while the pulmonary veins lie more peripherally. This part of the circulatory system is involved in gaseous exchange, and until birth little blood flows through the pulmonary vessels. For the changes to this circulatory system that occur at birth see Chapter 29.

Bronchial vessels supply the tissues of the lung. These vessels are initially direct branches from the paired dorsal aortae.

Clinical relevance

Respiratory distress syndrome (hyaline membrane disease) caused by a lack of surfactant results in atelectasis (lung collapse). This affects premature infants, and treatment options include a dose of steroids given to the infant to stimulate surfactant production, or surfactant therapy. Surfactant is administered to the infant directly down a tracheal tube. These treatments together with oxygen therapy and the application of a continuous positive airway pressure using a mechanical ventilator mean that the prognosis is good in many cases.

Oesophageal atresia and **tracheoeosphageal fistulas** are relatively common abnormalities. If the separation of the trachea from the foregut is incomplete various types of communicating passages may persist. This type of abnormality is often associated with other faults, including cardiac defects, limb defects and anal atresia. It is also possible that an oesophageal atresia will lead to polyhydramnios as the amniotic fluid is not swallowed by the foetus, or pneumonia after birth as fluid may enter the trachea through the fistula. Surgery is generally required.

Ectopic lung lobes and abnormalities in the branching of the bronchial tree rarely produce symptoms.

Congenital cysts of the lung can result in common infection sites and difficulty in breathing.

31 Digestive system: gastrointestinal tract

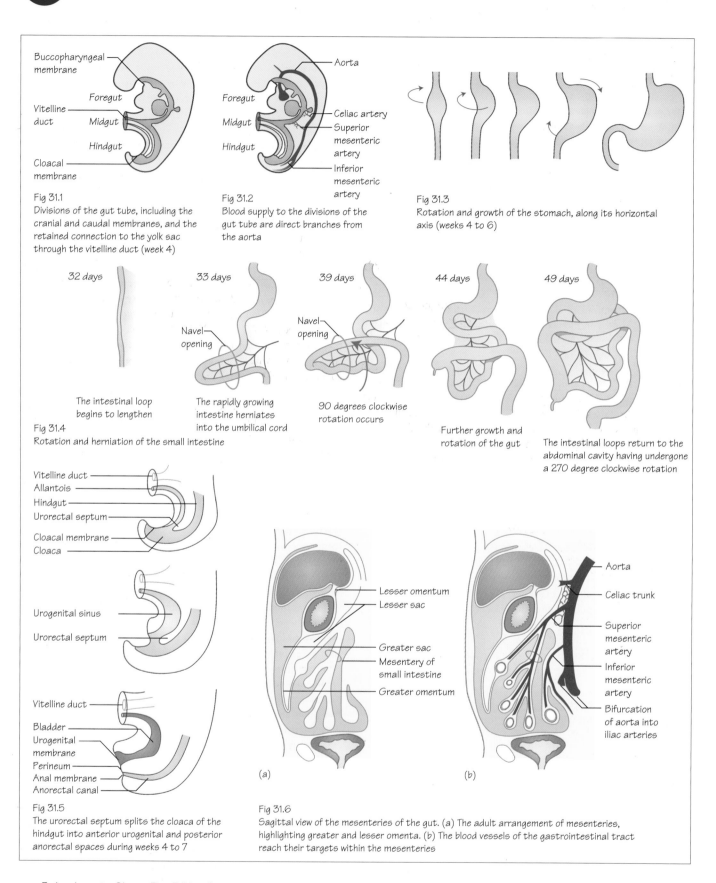

Fig 31.1
Divisions of the gut tube, including the cranial and caudal membranes, and the retained connection to the yolk sac through the vitelline duct (week 4)

Fig 31.2
Blood supply to the divisions of the gut tube are direct branches from the aorta

Fig 31.3
Rotation and growth of the stomach, along its horizontal axis (weeks 4 to 6)

32 days — The intestinal loop begins to lengthen

33 days — Navel opening — The rapidly growing intestine herniates into the umbilical cord

39 days — Navel opening — 90 degrees clockwise rotation occurs

44 days — Further growth and rotation of the gut

49 days — The intestinal loops return to the abdominal cavity having undergone a 270 degree clockwise rotation

Fig 31.4
Rotation and herniation of the small intestine

Fig 31.5
The urorectal septum splits the cloaca of the hindgut into anterior urogenital and posterior anorectal spaces during weeks 4 to 7

Fig 31.6
Sagittal view of the mesenteries of the gut. (a) The adult arrangement of mesenteries, highlighting greater and lesser omenta. (b) The blood vessels of the gastrointestinal tract reach their targets within the mesenteries

Embryology at a Glance, First Edition. Samuel Webster and Rhiannon de Wreede.

Time period: days 21–50

Induction of the tube

The gut tube forms when the yolk sac is pulled into the embryo and pinched off (see Figure 18.2) as the flat germ layers of the early embryo fold laterally and cephalocaudally (head to tail). Consequently, it has an endodermal lining throughout with a minor exception towards the caudal end. Epithelium forms from the endoderm layer and other structures are derived from the mesoderm.

Initially, the tube is closed at both ends, although the middle remains in contact with the yolk sac through the **vitelline duct** (or stalk) even as the yolk sac shrinks (Figure 31.1).

The cranial end will become the mouth and is sealed by the **buccopharyngeal membrane**, which will break in the fourth week, opening the gut tube to the amniotic cavity. The caudal end will become the anus and is sealed by the **cloacal membrane**, which will break during the seventh week.

Buds develop along the length of the tube that will form a variety of gastrointestinal and respiratory structures (see Chapter 32).

Divisions of the gut tube

The gut is divided into foregut, midgut and hindgut sections by the region of the gut tube that remains linked to the yolk sac and by the anterior branches from the aorta that supply blood to each part (Figure 31.2).

The foregut will develop into the pharynx, oesophagus, stomach and the first two parts of the duodenum to the major duodenal papilla, at which the common bile duct and pancreatic duct enter. The midgut includes the remainder of the duodenum and the small and large intestine through to the proximal two-thirds of the transverse colon. The hindgut includes the distal third of the transverse colon and the large intestine through to the upper part of the anal canal.

Blood supply

Each division of the gut is supplied by a different artery. The foregut is supplied by branches from the **coeliac artery** directly from the descending aorta. The midgut receives blood from the **superior mesenteric** artery and the hindgut from the **inferior mesenteric** artery (Figure 31.2).

Lower foregut

The foregut grows in length with the embryo, and epithelial cells proliferate to fill the lumen. The tube is later recanalised and only becomes a squamous epithelium during the foetal period. Failure of this normal process causes problems of stenosis (narrowing) or atresia (blocked) in the oesophagus or duodenum.

Part of the foregut tube begins to dilate in week 4, the dorsal side growing faster than the ventral side until week 6. This will become the stomach, and the dorsal side becomes the greater curvature. The dorsal mesentery (dorsal mesogastrium) will expand significantly to form the greater omentum.

The stomach rotates to bring the left side around to become the ventral surface, explaining why the left vagus nerve innervates the anterior of the stomach (Figure 31.3). This rotation also moves the duodenum into the adult C-shaped position.

Twists of the midgut

The midgut also lengthens considerably, looping and twisting as it does so, filling the abdominal cavity. At approximately 6 weeks the midgut grows so quickly there is not enough room in the abdomen to contain it, and it herniates into the umbilical cord (Figure 31.4).

The midgut also rotates through **270° counterclockwise** (if you were to be looking at the abdomen), bringing the developing caecum from the inferior abdomen up the left of the developing small intestine to the top of the abdomen, and around to descend to its adult location in the lower right quadrant. The axis of this rotation is the superior mesenteric artery and the rotation is of particular significance when considering the layout of the small and large intestines and accessory organs in adult anatomy.

The midgut re-enters the abdomen in week 10, and it is thought that growth of the abdomen together with regression of the mesonephric kidney and a reduced rate of liver growth are important factors in this occurring normally.

Story of the hindgut and the cloaca

The last part of the gut tube, the hindgut, ends initially in a simple cavity called the cloaca. The cloaca is also continuous with the allantois, a remnant of the yolk sac that largely regresses but contributes to the superior parts of the bladder in the human embryo.

A wedge of mesoderm, the **urorectal septum**, moves caudally towards the cloacal membrane as the embryo grows and folds during weeks 4–7 (Figure 31.5). The urorectal septum divides the cloaca into a primitive **urogenital sinus** anteriorly and an **anorectal canal** posteriorly. The urogenital sinus will form parts of the bladder and the urogenital tract.

The cloacal membrane ruptures in the seventh week, opening the gut tube to the amniotic cavity. The caudal part of the lining of the anal canal is thus derived from ectoderm and the cephalic part from endoderm. Subsequently, the caudal part of the anal canal receives blood from branches of the internal iliac arteries and the cephalic part receives blood from the artery of the hindgut, the inferior mesenteric artery. Similarly, portosystemic anastomoses also occur here.

Mesenteries

Mesenteries of the gut form as a covering of mesenchyme passing over the gut tube from the posterior body wall of the embryo when the tube is in close contact with it. With growth the gut tube moves further into the abdominal cavity and away from the posterior wall. A bridging connective tissue forms suspending the gut and its associated organs within the abdomen in a dorsal mesentery for most of its length and a ventral mesentery around the lower foregut region. The ventral mesentery is derived from the septum transversum.

The dorsal mesentery will form the mesenteries of the small and large intestines of the adult gastrointestinal tract, and also forms the **greater omentum** (Figure 31.6). The ventral mesentery will form the **lesser omentum** between the stomach and the liver, and the falciform ligament between the liver and the anterior abdominal wall.

The extensive lengthening and rotation of the midgut causes the dorsal mesentery to become considerably larger and more convoluted, and its initial simplicity explains the short diagonal attachment of the mesentery of the small intestine to the posterior abdominal wall in the adult. When the hindgut finds its final position in the foetus the mesenteries of the ascending and descending colon fuse with the peritoneum of the posterior body wall.

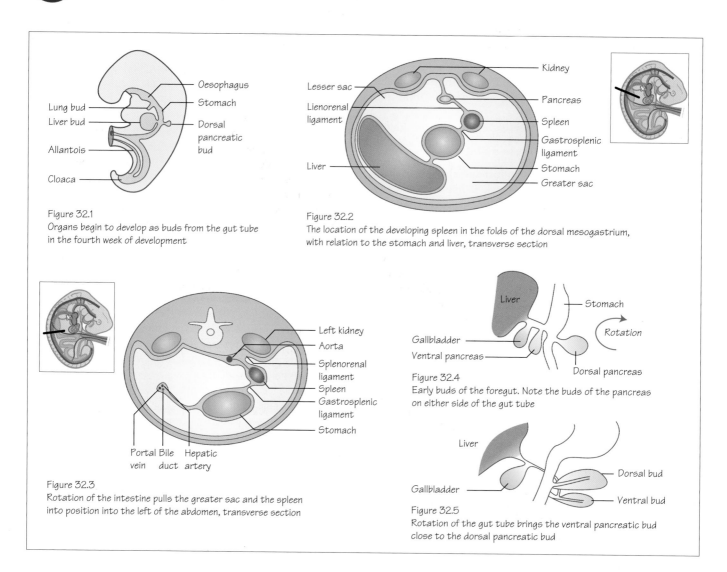

Figure 32.1
Organs begin to develop as buds from the gut tube in the fourth week of development

Figure 32.2
The location of the developing spleen in the folds of the dorsal mesogastrium, with relation to the stomach and liver, transverse section

Figure 32.3
Rotation of the intestine pulls the greater sac and the spleen into position into the left of the abdomen, transverse section

Figure 32.4
Early buds of the foregut. Note the buds of the pancreas on either side of the gut tube

Figure 32.5
Rotation of the gut tube brings the ventral pancreatic bud close to the dorsal pancreatic bud

Embryology at a Glance, First Edition. Samuel Webster and Rhiannon de Wreede.

Time period: day 21 to birth

Introduction

In Chapter 31 we looked at the development of the gastrointestinal tract as a tube and mentioned a number of buds that sprout from the tube and its associated mesenchyme. These develop into a number of organs (Figure 32.1).

Lung bud

As the oesophagus develops and elongates during week 4 the **respiratory diverticulum** buds off from its ventral wall (Figure 32.1). To create two separate tubes a septum forms between the respiratory bud and the oesophagus called the **tracheoesophageal septum** (see Figure 30.1). This creates the oesophagus dorsally and the respiratory primordium ventrally (see Chapter 30).

Spleen

In the fifth week the spleen starts to develop from a condensation of mesenchymal cells between the folds of the dorsal mesogastrium (Figure 32.2). With the rotation of the stomach and duodenum the spleen is moved to the left side of the abdomen, explaining the adult location of the splenic artery, a branch of the coeliac trunk. The **gastrosplenic ligament** between the stomach and spleen is an adult remnant of the dorsal mesogastrium, as is the **splenorenal ligament** between the spleen and left kidney (Figure 32.3).

The spleen begins to create red and white blood cells in the second trimester and is an important site of haematopoesis during the foetal period. After birth it stops producing red blood cells and concentrates on its adult functions of the lymphatic and immune systems, and of removing old red blood cells from circulation.

Liver and gallbladder

Beginning as an epithelial outgrowth from the ventral wall of the distal end of the foregut the **liver bud**, or **hepatic diverticulum** (Figure 32.1), appears at the end of week 3. Growing rapidly during week 4 the liver bud grows into the **septum transversum**, a sheet of mesodermal cells located between the pericardial cavity and the yolk sac stalk. The septum transversum will contribute to the diaphragm (see Chapter 17) and the ventral mesentery here. Both the liver bud and septum transversum integrate to form parts of the liver. The liver bud grows within the ventral mesentery, and

retains a connection with the foregut that will become the **bile duct**. A cranial part of the liver bud will form the liver, and a caudal bud will form the **gallbladder** (Figure 32.4).

The liver is formed from cells of different sources. The liver bud from the foregut will form hepatocytes and the epithelial lining of the bile duct. The vitelline and umbilical veins will form hepatic sinusoids. Cells of the septum transversum will form the stroma and capsule (connective tissues) of the liver and also haematopoietic cells, Kupffer cells, smooth muscle and connective tissue of the biliary tract. The lesser omentum between the stomach and liver, and the falciform ligament between the liver and the anterior abdominal wall are the adult structures of the ventral mesentery.

By week 10 of development the liver accounts for around 10% of the embryonic weight. At birth this reduces to 5% of total body weight. A main embryological function of the liver is haematopoiesis, with the liver producing red and white blood cells.

With the rotation of the stomach and duodenum the route of the **common bile duct** to the duodenum is altered from anterior to the foregut to a posterior course (Figure 32.5), and is joined by the pancreatic duct at the **ampulla of Vater**. Eventually the bile duct passes behind the duodenum and bile is formed by the liver in week 12.

Pancreas

Two pancreatic buds develop from the foregut (duodenum) giving dorsal and ventral buds (in the fourth and fifth week, respectively) within the mesentery. The dorsal bud is larger, and the ventral bud is a bud from the hepatic diverticulum (Figure 32.4).

With the rotation of the duodenum to the right the ventral bud moves dorsally (much like the movement of the bile duct entrance to the duodenum) to rest below and behind the dorsal bud (Figure 32.5). In week 7 the duct systems of the buds fuse and the adult main pancreatic duct forms from the main duct of the ventral bud and the distal part from the dorsal bud. Occasionally, the proximal part of the duct of the dorsal bud persists as an accessory duct that opens into the duodenum a little proximal to the main duct.

The uncinate process and most of the head of the pancreas forms from the ventral bud, and the rest forms from the dorsal bud. Exocrine and endocrine cells are all derived from endoderm, taking separate differentiation pathways. The **islets of Langerhans** (endocrine cells) form in the third month and insulin is secreted from the fourth to fifth month.

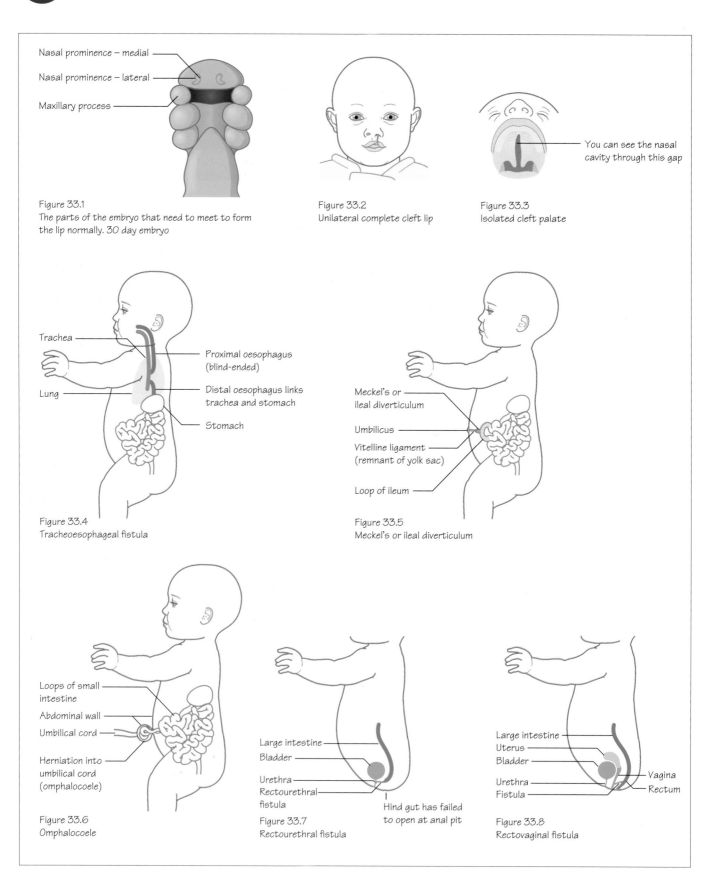

Figure 33.1
The parts of the embryo that need to meet to form the lip normally. 30 day embryo

Nasal prominence – medial
Nasal prominence – lateral
Maxillary process

Figure 33.2
Unilateral complete cleft lip

Figure 33.3
Isolated cleft palate

You can see the nasal cavity through this gap

Figure 33.4
Tracheoesophageal fistula

Trachea
Lung
Proximal oesophagus (blind-ended)
Distal oesophagus links trachea and stomach
Stomach

Figure 33.5
Meckel's or ileal diverticulum

Meckel's or ileal diverticulum
Umbilicus
Vitelline ligament (remnant of yolk sac)
Loop of ileum

Figure 33.6
Omphalocoele

Loops of small intestine
Abdominal wall
Umbilical cord
Herniation into umbilical cord (omphalocoele)

Figure 33.7
Rectourethral fistula

Large intestine
Bladder
Urethra
Rectourethral fistula
Hind gut has failed to open at anal pit

Figure 33.8
Rectovaginal fistula

Large intestine
Uterus
Bladder
Urethra
Fistula
Vagina
Rectum

Embryology at a Glance, First Edition. Samuel Webster and Rhiannon de Wreede.

Time period: birth

Facial abnormalities

A relatively common congenital abnormality is **cleft lip** and/or **cleft palate** which affects around 1 in 600–700 live births and has a collection of defects.

Cleft lip (*cheiloschisis*) can be incomplete (affects upper lip only) or complete (continues into the nose) and unilateral (Figure 33.1) or bilateral. It is caused by the incomplete fusion of the medial nasal prominence with the maxillary process (Figure 33.2). When these fuse normally they form the intermaxillary segment, which goes on to become the primary (soft) palate.

The secondary (hard) palate forms from outgrowths of the maxillary process called the palatine shelves. Failure of these shelves to fuse or ascend to a horizontal position causes **cleft palate** (*palatoschisis*). In very severe cases the cleft can continue into the upper jaw. Cleft palate is often accompanied by cleft lip (complete), but not always (incomplete; Figure 33.3), and can also be unilateral or bilateral.

A cleft lip is generally diagnosed at the 20-week anomaly scan, whereas cleft palates are diagnosed after birth. Cleft lips require surgical intervention before 3 months, whereas cleft palate surgery should happen before the child reaches 12 months old. Cleft lip and palate can affect feeding and speech, but also hearing. To aid prevention of cleft lip and palate maternal dietary folic acid is recommended (see also spina bifida, Chapter 15).

Foregut abnormalities

Abnormalities in development of the foregut can include stenosis and atresia at various points along its length, and hypertrophy of the pylorus of the stomach. Depending upon the point of restriction projectile vomiting can be a symptom, and the presence or absence of bile in the vomit can help diagnose the location.

The respiratory tract forms as a bud from the foregut, so a **tracheoesophageal fistula** can form (Figure 33.4). The most common variant sees the proximal oesophagus end blindly and the trachea connected to the distal oesophagus. There are many other variations and frothy oral secretions are often a symptom. Surgery is required.

A **congenital hiatal hernia** is caused by the oesophagus not lengthening fully, preventing the diaphragm from forming normally and pulling the top of the stomach up into the thorax. This can affect the development of respiratory structures, and occurs in varying severity.

Midgut abnormalities

A remnant of the vitelline duct that connected the yolk sac to the midgut may persist as an ileal diverticulum (also known as Meckel's diverticulum; Figure 33.5) or as a **vitelline cyst** (also known as an omphalomesenteric duct cyst) in the distal ileum. An ileal diverticulum is present in around 2% of the population, but the majority are asymptomatic. Ulceration may form here with bleeding. If the vitelline duct persists as fibrous cords between the abdominal wall and the ileum loops of intestine may become twisted around it. The duct may survive as a true duct between the ileum and the external umbilicus.

The midgut may fail to complete its rotation or to fail to rotate in the normal direction during development, giving **abnormal rotation** or **reverse rotation of intestine**. Abnormal rotation is caused by only a 90° rotation and gives a left-sided colon, whereas reverse rotation causes the transverse colon to lie posterior to the superior mesenteric artery after a 90° clockwise rotation of the midgut instead of the normal 270° counterclockwise rotation.

Omphalocoele (or exomphalos) is the herniation of abdominal contents into the umbilicus, and the contents remain covered by peritoneum and amnion (Figure 33.6). This can normally be diagnosed by antenatal ultrasound scanning. Omphalocoele is thought to occur as a failure of the midgut to reenter the abdominal cavity after the normal herniation of weeks 6–10. Omphalocoele is often associated with cardiac and neural tube defects, trisomy 13 and 18 and Beckwith–Wiedemann syndrome.

Hindgut abnormalities

The urorectal septum normally separates the cloaca into urogenital and hindgut spaces. If this fails to occur normally links between the two spaces can occur, such as a **rectourethral (urorectal) fistula** (Figure 33.7) or a **rectovaginal fistula** (Figure 33.8).

Anal atresia can occur, possibly as a result of interrupted blood supply during development. **Imperforate anus** may also occur if the cloacal membrane does not break down. There are different degrees of severity, and some require a colostomy whereas others are repairable with surgical intervention, often within 24 hours of birth.

Associated organs
Liver

Jaundice affects 60% of healthy newborn infants and has multiple causes, often categorised by age of onset. It is normally identified through the infant's skin colour and bilirubin levels. Most cases of jaundice do not need treatment, but phototherapy helps reduce bilirubin levels. In extreme cases an exchange transfusion is necessary.

Pancreas

Due to abnormalities in the rotation of the ventral bud pancreatic tissue can end up surrounding the duodenum. This is called an **annular pancreas**. It is possible that this tissue can constrict the duodenum and cause a complete blockage. Early signs can include **polyhydramnios**. It is normally treated with surgery.

Spleen

Splenic lobulation and an accessory spleen are relatively common. Rarer conditions include a wandering spleen and **polysplenia** (multiple accessory spleens).

Splenogonadal fusion, a very rare developmental anomaly, results from the abnormal fusion of the splenic and gonadal primordia during prenatal development.

Hyposplenism (reduced splenic function) may occur because of a congenital failure of the spleen to form. Affected individuals are at increased risk of bacterial sepsis.

34 Urinary system

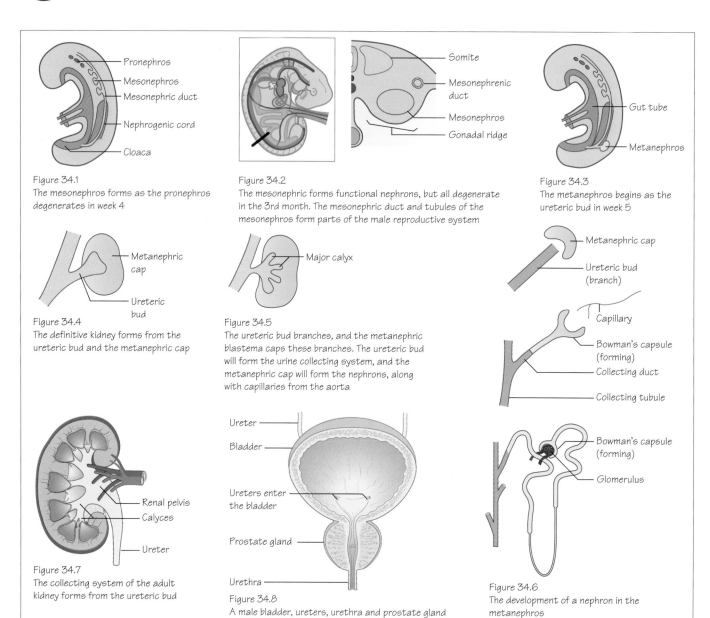

Figure 34.1
The mesonephros forms as the pronephros degenerates in week 4

(labels: Pronephros, Mesonephros, Mesonephric duct, Nephrogenic cord, Cloaca)

Figure 34.2
The mesonephric forms functional nephrons, but all degenerate in the 3rd month. The mesonephric duct and tubules of the mesonephros form parts of the male reproductive system

(labels: Somite, Mesonephrenic duct, Mesonephros, Gonadal ridge)

Figure 34.3
The metanephros begins as the ureteric bud in week 5

(labels: Gut tube, Metanephros)

Figure 34.4
The definitive kidney forms from the ureteric bud and the metanephric cap

(labels: Metanephric cap, Ureteric bud)

Figure 34.5
The ureteric bud branches, and the metanephric blastema caps these branches. The ureteric bud will form the urine collecting system, and the metanephric cap will form the nephrons, along with capillaries from the aorta

(labels: Major calyx)

(labels: Metanephric cap, Ureteric bud (branch), Capillary, Bowman's capsule (forming), Collecting duct, Collecting tubule, Bowman's capsule (forming), Glomerulus)

Figure 34.7
The collecting system of the adult kidney forms from the ureteric bud

(labels: Renal pelvis, Calyces, Ureter)

Figure 34.8
A male bladder, ureters, urethra and prostate gland

(labels: Ureter, Bladder, Ureters enter the bladder, Prostate gland, Urethra)

Figure 34.6
The development of a nephron in the metanephros

Time period: day 21 to birth

Introduction
The development of the urinary system is closely linked with that of the reproductive system. They both develop from the **intermediate mesoderm,** which extends on either side of the aorta and forms a condensation of cells in the abdomen called the **urogenital ridge**.

The ridge has two parts: the **nephrogenic cord** and the **gonadal ridge** (see Figure 36.2).

Kidneys
Three structures involved in kidney development grow from intermediate mesoderm in an anterior to posterior sequence, termed the **pronephros, mesonephros** and **metanephros**.

Embryology at a Glance, First Edition. Samuel Webster and Rhiannon de Wreede.

The **pronephros** appears in the third week in the neck region of the embryo and disappears a week later. In humans this is a primitive, non-functional kidney that consists of vestigial nephrons joined to an unbranched nephric duct.

Mesonephros

Appearing in the fourth week the first functional kidney unit, the mesonephros, forms as the pronephros begins to regress (Figures 34.1 and 34.2). The **mesonephric ducts (Wolffian ducts)** are epithelia-lined tubes that form in the intermediate mesoderm and extend caudally to the cloaca. They stimulate formation of the mesonephros itself as **mesonephric tubules** (different to the ducts) from the mesenchyme. The tissue of the mesonephros appears initially as a segmented structure along the mesonephric duct.

Renal corpuscles develop from mesonephric tubules (**Bowman's capsule**) and capillaries from the dorsal aortae (**glomerulus**). At the lateral end the tubules join the mesonephric duct. The duct discharges into the cloaca where the bladder will form. The mesonephros starts to produce urine at about 6 weeks but degenerates almost completely between weeks 7 and 10.

The mesonephric ducts contribute to the ducts of the male reproductive system, but regress in the female foetus (see Chapter 36).

Metanephros

The third renal structure that develops will finally become the adult kidney. It starts to appear at the beginning of the fifth week as a bud from the caudal end of the mesonephric duct, called the **ureteric bud** (Figures 34.3 and 34.4).

The bud branches and develops into the collecting parts of the adult kidney: the ureter, renal pelvis, calyses and collecting tubules. The bud grows into surrounding intermediate mesoderm and induces the cells in that region (the metanephric blastema) to form a **metanephric cap** upon the ureteric bud.

As the ureteric bud forms collecting tubules, cells of the metanephric cap form **nephrons** that link to the collecting tubules. Reciprocal interactions between the buds and the caps initiate and maintain this development (Figure 34.5).

Capillaries grow into the Bowman's capsule from the dorsal aortae and convolute to form the glomeruli (Figure 34.6). These functional renal units produce urine from week 12 onwards.

The formation of nephrons continues until birth when there are approximately 1 million nephrons in each kidney. Infant kidneys are lobulated because of the branching of the calyces (Figure 34.7), but further growth and elongation of the nephrons after birth pushes out the kidney and the lobulation disappears.

Blood supply

The location of the metanephros changes during development from the level of the pelvis, through growth of the embryo and migration of the kidneys, to the lumbar region. They also rotate medially in ascent. As they ascend, a series of blood vessels from either the common iliac arteries or aorta generate and degenerate to continually supply the kidneys. Usually, the most cranial remain and become the renal arteries.

Bladder and urethra

In week 4 the **cloaca** is split into the ventral **urogenital sinus** and the dorsal **anal canal** by the urorectal septum (see Figure 31.5).

The urogenital sinus can be split into a further three parts. The top part is the biggest and becomes the bladder, the middle part forms the urethra in the female pelvis and the prostatic and membranous urethra in the male (Figure 34.8), and the lowest part forms the penile urethra in the male and the vestibule in the female. The allantois also contributes to the upper parts of the bladder.

The mesonephric ducts become incorporated into the posterior wall of the bladder. The openings of the mesonephric ducts and ureters enter the bladder separately. Remember that the ureters form from the metanephric ducts. The ureters move anteriorly whereas the mesonephric ducts move posteriorly and become the ejaculatory ducts in the male pelvis.

The specialised **transitional epithelium** of the bladder develops from the endoderm of the urogenital sinus.

The ventral surface of the cloaca (which becomes the urogenital sinus) is continuous with the allantois, which degenerates after birth to form the **urachus** and eventually the **median umbilical ligament** (an embryological remnant with no clinical significance). The *medial* umbilical ligaments are the remnants of the umbilical arteries, which are a little lateral to the urachus.

Clinical relevance

Incomplete division of the ureteric bud can lead to **supernumerary kidneys** and, more commonly, **supernumerary ureters**.

Kidney cysts form when the developing nephrons fail to connect to a collecting tubule in development, or the collecting ducts fail to develop. There are dominant and recessive forms of **polycystic kidneys**. The recessive form is more progressive and often results in renal failure in childhood.

Balance of fluid in the amnion is vital in the development of the embryo. If urine is not being produced there is a reduction in the amniotic fluid and **oligohydramnios** develops. This can be a symptom of bilateral **renal agenesis**, in which both kidneys fail to form. This is lethal. Unilateral renal agenesis generally causes no symptoms.

Accessory renal arteries are quite common, especially on the left and often are only seen during a surgical procedure as they are asymptomatic. They enter the kidney at the superior and inferior poles. Abnormal rotation or location of the kidneys may be found in a patient, and they may fail to ascend into the abdomen. The inferior poles of the left and right kidneys can fuse, forming a horseshoe kidney. In this case the kidney cannot ascend as it gets snagged on the inferior mesenteric artery.

Bladder defects may occur, such as **exstrophy** in which part of the ventral bladder wall is present outside of the abdominal wall. A **urachal cyst, fistula** or **sinus** can form if the degeneration of the allantois is not completed.

35 Reproductive system: ducts and genitalia

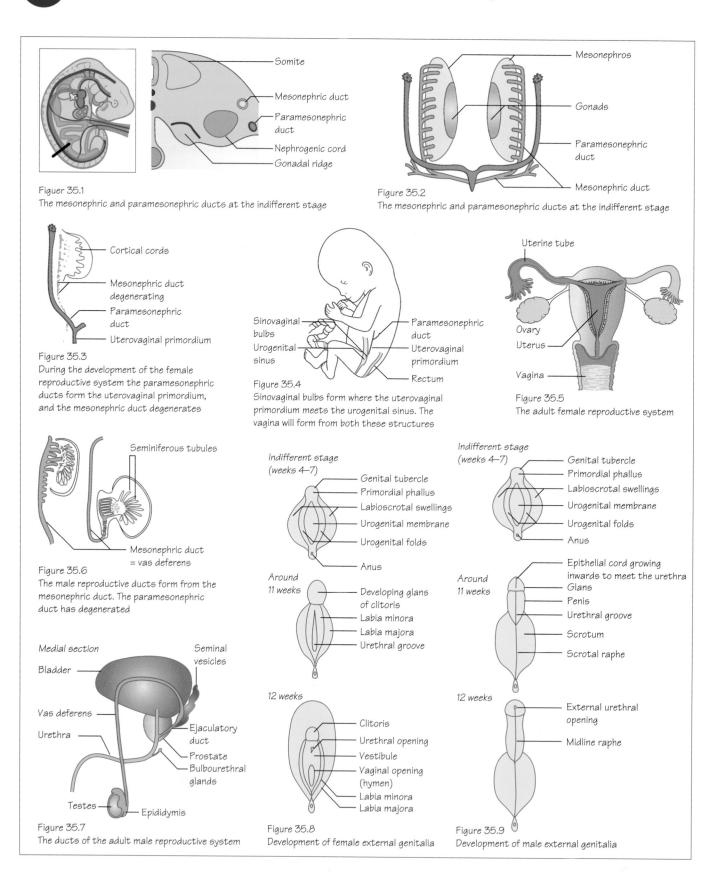

Figuer 35.1
The mesonephric and paramesonephric ducts at the indifferent stage

Labels (Figure 35.1): Somite, Mesonephric duct, Paramesonephric duct, Nephrogenic cord, Gonadal ridge

Figure 35.2
The mesonephric and paramesonephric ducts at the indifferent stage

Labels (Figure 35.2): Mesonephros, Gonads, Paramesonephric duct, Mesonephric duct

Figure 35.3
During the development of the female reproductive system the paramesonephric ducts form the uterovaginal primordium, and the mesonephric duct degenerates

Labels (Figure 35.3): Cortical cords, Mesonephric duct degenerating, Paramesonephric duct, Uterovaginal primordium

Figure 35.4
Sinovaginal bulbs form where the uterovaginal primordium meets the urogenital sinus. The vagina will form from both these structures

Labels (Figure 35.4): Sinovaginal bulbs, Urogenital sinus, Paramesonephric duct, Uterovaginal primordium, Rectum

Figure 35.5
The adult female reproductive system

Labels (Figure 35.5): Uterine tube, Ovary, Uterus, Vagina

Figure 35.6
The male reproductive ducts form from the mesonephric duct. The paramesonephric duct has degenerated

Labels (Figure 35.6): Seminiferous tubules, Mesonephric duct = vas deferens

Figure 35.7
The ducts of the adult male reproductive system

Labels (Figure 35.7): Medial section, Bladder, Vas deferens, Urethra, Testes, Seminal vesicles, Ejaculatory duct, Prostate, Bulbourethral glands, Epididymis

Figure 35.8
Development of female external genitalia

Labels (Figure 35.8): Indifferent stage (weeks 4–7), Genital tubercle, Primordial phallus, Labioscrotal swellings, Urogenital membrane, Urogenital folds, Anus; Around 11 weeks, Developing glans of clitoris, Labia minora, Labia majora, Urethral groove; 12 weeks, Clitoris, Urethral opening, Vestibule, Vaginal opening (hymen), Labia minora, Labia majora

Figure 35.9
Development of male external genitalia

Labels (Figure 35.9): Indifferent stage (weeks 4–7), Genital tubercle, Primordial phallus, Labioscrotal swellings, Urogenital membrane, Urogenital folds, Anus; Around 11 weeks, Epithelial cord growing inwards to meet the urethra, Glans, Penis, Urethral groove, Scrotum, Scrotal raphe; 12 weeks, External urethral opening, Midline raphe

Embryology at a Glance, First Edition. Samuel Webster and Rhiannon de Wreede.

Time period: day 35 to postnatal development

Introduction

The reproductive systems develop from a series of epithelial cell-lined ducts, derived from mesoderm. The initial stage of genital development is the same for both sexes up to week 7, and is called the **indifferent stage**.

Ducts

The indifferent stage involves the **mesonephric ducts** (or Wolffian ducts) from the developing urinary system and the **paramesonephric ducts** (or Müllerian ducts), named because of their location lateral to the mesonephric ducts (Figures 35.1 and 35.2). The paramesonephric ducts form from longitudinal invaginations of the surface epithelium of the gondal ridge.

Female

The paramesonephric ducts descend, meet in the midline and fuse in the pelvic region to form the **uterovaginal primordium** (Figure 35.3). This bulges into the dorsal wall of the developing urogenital sinus (see Chapters 31 and 34) but does not break the wall. The bulge forms the **paramesonephric tubercle** (or sinus tubercle, or Müller tubercle).

The paramesonephric ducts open into the peritoneal cavity, and the free unfused cranial ends become the **uterine tubes**. The **uterus** forms from the midline uterovaginal primordium.

The paramesonephric tubercle induces the urogenital sinus to form 2 outgrowths of cells within its lumen. These outgrowths proliferate and form the **sinovaginal bulbs**, which fuse and form the **vaginal plate** (Figure 35.4). This will canalise to form a hollow core, which is completed by the fifth month.

The inferior part of the vagina probably forms from the vaginal plate, and the superior part from uterovaginal primordium. The vagina is separated from the urogenital sinus by the hymen.

The female reproductive system (Figure 35.5) is likely to grow from 2 tissue origins: the lining of the lower portion of the vagina is endodermal and the upper portion, fornices and uterus are mesodermal. The muscle and connective tissues of the vagina and uterus are derived from the surrounding mesenchyme.

The mesonephric ducts degenerate, although remnants may remain.

Male

Mesonephric ducts become the efferent ductules and epididymis of the testes, the ductus deferens (or vas deferens) and the ejaculatory duct (Figures 35.6 and 35.7).

The seminal vesicles form as an outgrowth from the ductus deferens, whereas the prostate gland arises from numerous outgrowths from the urethra. The endodermal cells of the urethra differentiate to become the glandular tissue of the prostate gland, and the surrounding mesenchyme forms the smooth muscle and connective tissue.

Paramesonephric ducts degenerate (although remnants can remain).

External genitalia

Until the ninth week of development the external genitals appear the same for both sexes (Figures 35.8 and 35.9). You cannot see the difference in the sex of a developing embryo until around 11 weeks' gestation. To prevent mistakes made in ultrasound identification, if the sex of the foetus is required it is identified at the 20-week scan.

During the indifferent stage, the **cloacal membrane** is surrounded by mesenchymal folds called **urogenital** (cloacal) **folds** that fuse ventrally into a **genital tubercle**. Around week 7, the **urogenital septum** splits the cloacal membrane into a ventral **urogenital membrane** and a dorsal **anal membrane**.

Another pair of folds develop lateral to the urogenital folds, called the **labioscrotal swellings**. The urogenital membrane degenerates leaving the urogenital sinus in direct communication with the amniotic cavity. The genital tubercle elongates and forms the **primordial phallus**.

Female

Induced by **oestrogens** secreted from the placenta and foetal ovaries, the genital turbercle develops into the **clitoris** (Figure 35.8). During the third and fourth months the clitoris is larger than its male counterpart. The urogenital groove remains open and develops into the **vestibule** which contains the openings of the vagina and urethra. The urogenital folds remain largely unfused (the two sides only meet posteriorly) and become the **labia minora**. The labioscrotal swellings become the **labia majora**.

Male

Induced by androgens secreted from the developing testes, the primordial phallus grows to form the **penis** (Figure 35.9). The urogenital sinus forms a groove bound laterally by the urogenital folds, and endodermal cells divide and line the groove which is now termed the **urethral plate**. The urethral folds eventually fuse on the underside (penile raphe) surrounding a tube (the spongy part of the urethra).

The urethra temporarily ends blindly in the anterior part of the penis. In the fourth month the terminal part of the urethra is formed when cells from the glans grow internally producing an epithelial cord. A lumen then forms and creates the external urethral meatus. The lateral genital swellings form the **scrotum** and the visible line of fusion is the scrotal raphe.

Sex determination

The SRY gene (sex-determining region of the Y chromosome) encodes for a transcription factor that is expressed in the gonad during the indifferent stage, triggering male development. If this transcription factor is absent female development occurs.

Clinical relevance

Hypospadias is caused by incomplete fusion of the urethral folds in the male, and the urethra opens onto the ventral surface of the penis. **Epispadias** results from the genital tubercle developing in the area of the urorectal septum, causing the urethra to open on the dorsal surface of the penis. Epispadias usually occurs in males but can occur in females and results in a split clitoris and an abnormally positioned urethral opening.

Congenital adrenal hyperplasia is an enzyme deficiency causing the adrenal glands to fail to produce sufficient cortisol and aldosterone, but the body produces excess androgens. This can result in ambiguous genitalia development in females but will not affect males. Further developmental problems occur, such as precocious puberty.

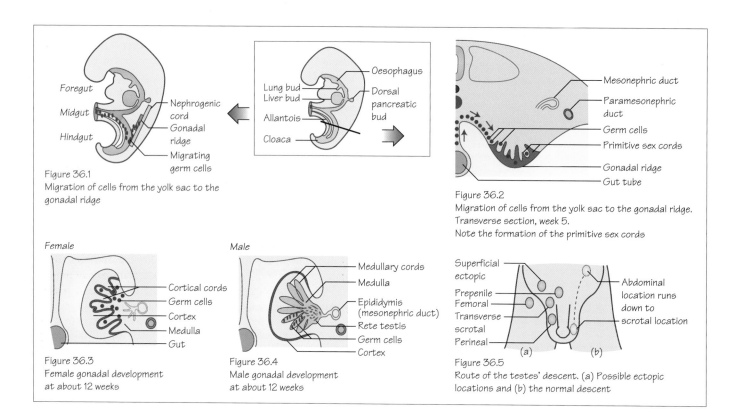

Figure 36.1
Migration of cells from the yolk sac to the gonadal ridge

Figure 36.2
Migration of cells from the yolk sac to the gonadal ridge.
Transverse section, week 5.
Note the formation of the primitive sex cords

Figure 36.3
Female gonadal development
at about 12 weeks

Figure 36.4
Male gonadal development
at about 12 weeks

Figure 36.5
Route of the testes' descent. (a) Possible ectopic
locations and (b) the normal descent

Embryology at a Glance, First Edition. Samuel Webster and Rhiannon de Wreede.

80 © 2012 John Wiley & Sons, Ltd. Published 2012 by John Wiley & Sons, Ltd.

Time period: day 30 to postnatal development

Introduction
In the chapter on renal development (see Chapter 34) we talked about the development of the **gonadal ridge** from **intermediate mesoderm**, an important source of cells for the reproductive system and the location for the beginning of the development of the gonads.

Gonads
Gonads are formed from three sources of cells: the **intermediate mesoderm**, the **mesodermal epithelium** that lines the developing urogenital ridge and **germ cells**.

Germ cells originate in the extra-embryonic endoderm of the yolk sac near the allantois and migrate along the dorsal mesentery of the hindgut to reach the gonadal ridge at the beginning of week 5 (Figure 36.1). By the sixth week they invade the **gonadal ridge** (see Figure 34.2). Also at this time the epithelium overlying the mesoderm begins to proliferate, penetrating the mesoderm and forming **cords** that are continuous with the surface epithelium (Figure 36.2).

This indifferent gonad has a discernible external **cortex** and internal **medulla**. If the migrating germ cells fail to arrive the gonads will not develop because of the absence of reciprocal interactions between germ cells and surrounding epithelia.

Female
In the early female gonad the cortex develops and the medulla regresses. The **primitive sex cords** dissociate and form irregular cell clusters containing germ cells (Figure 36.3). These cords and clusters disappear and are replaced with blood vessels and connective tissue.

Surface epithelia continue to proliferate and produce a second wave of sex cords that remain close to the surface. In the fourth month of development these also dissociate and form cell clusters surrounding one or more germ cells. This is the **primitive follicle** and the surrounding epithelial cells develop into follicular cells (see Figure 8.1). Each primitive germ cell becomes an **oogonium**. Oogonia divide significantly before birth but there is no division postnatally.

A part of peritoneum attached to the gonad develops into the **gubernaculum**. This structure passes through the abdominal wall (the future inguinal canal) and attaches to the internal surface of the labioscrotal swellings (see Figure 35.8). The ovaries descend into the pelvis, and the gubernaculum becomes attached to the uterus. In the adult the gubernaculum remains as the **round ligament** (passing through the inguinal canal) of the uterus and the **ovarian ligament**.

Male
The cortex regresses and the medulla develops (Figure 36.4). Testes develop quicker than ovaries, and the primitive sex cords do not degenerate but continue to grow into the medulla.

Testosterone producing cells, called **Leydig** cells, develop from mesoderm of the gonadal ridge and are located between the developing sex cords. They produce testosterone by week 8.

The primitive sex cords break up and form two networks of tubes: the **rete testis** and the **seminiferous tubules**. The **tunica albuginea** (thick fibrous connective tissue) develops to separate the networks from the surface epithelia. The rete testes are the connection between the seminiferous tubules and the efferent ducts of the testes (see Figure 7.1), which are derived from the **mesonephric tubules** (see Chapter 34).

In the fourth month the seminiferous tubules contain two important cell types: primitive germ cells that form spermatogonia, and **Sertoli cells** that have support roles for the cells passing through spermatogenesis. The male **gubernaculum** runs from the inferior pole of the testis to the labioscrotal folds (see Figure 35.9) and guides the testis into the scrotum, along with the ductus deferens and its blood vessels, as the foetus becomes longer and the pelvis becomes larger. The inguinal canal normally closes behind the testis, but failure of this process increases the risk of an indirect inguinal hernia.

Blood supply
The gonads develop in the abdomen and hence receive their blood supply directly from the abdominal aorta. The male arteries are called the **testicular arteries** and the female arteries the **ovarian arteries**. During the descent of the testes their blood vessels are pulled behind them as they pass through the inguinal canal and into the scrotum. The lymphatic system of both gonads also follows these pathways.

Clinical relevance
Undescended testes (cryptorchidism) describes the failure of the testes to descend normally into the scrotum by birth. This may occur bilaterally or unilaterally, and is more common in premature males. The testes may remain in the abdominal cavity, at a point along their normal route of descent or within the inguinal canal (Figure 36.5). Often, the testes will have descended to the scrotum by the end of the first year, but testes that remain undescended are likely to cause fertility problems. Undescended testes, even if they later descend, are linked to an increased risk of testicular cancer.

Hormonal imbalances can result in a varied range of developmental abnormalities to the reproductive system. Chromosomal defects are also responsible for many genital abnormalities, often presenting with other congenital defects. Those with **gonadal dysgenesis** have male chromosomes but no testes. Patients can have female external genitalia and underdeveloped female internal genitalia or ambiguous external genitalia and a mixture of both sexes internally, but are often raised as girls.

Ovarian and **testicular cancers** are relatively common forms of cancer. If **testicular cancer** is suspected it is often from a lump found in one testis and diagnosed through an ultrasound scan. It is important to remember that lymph drainage is to the retroperitoneal para-aortic lymph nodes rather than pelvic nodes, and these are involved in the staging of testicular cancer. Affected nodes must also be removed surgically together with the testis. The prognosis for testicular cancers is generally good. **Ovarian cancer** symptoms are often absent and if present, unspecific. An increase in abdominal size and urinary problems are possible. Surgical treatment is often required but because of the lack of early symptoms and diagnosis the prognosis is generally poor.

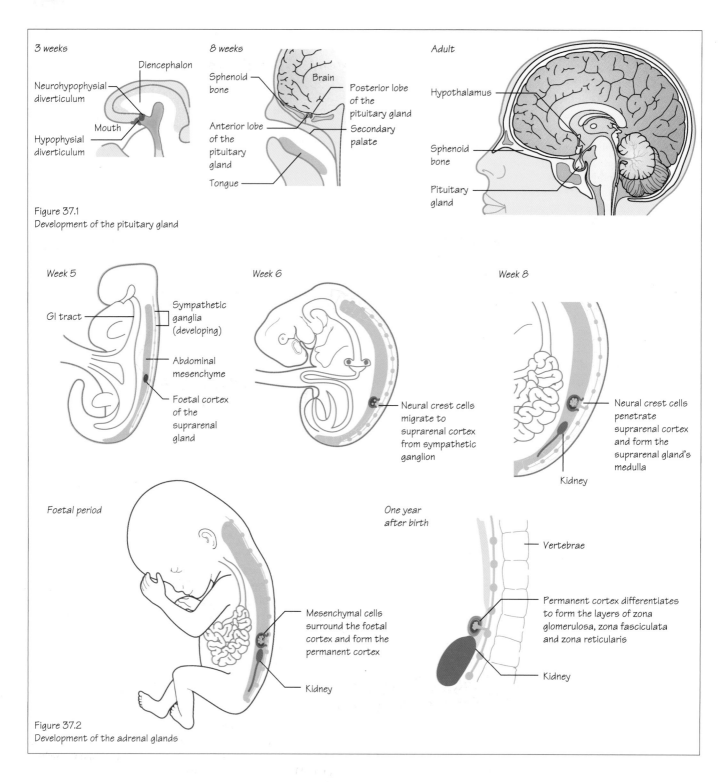

Figure 37.1
Development of the pituitary gland

3 weeks

Neurohypophysial diverticulum
Diencephalon
Mouth
Hypophysial diverticulum

8 weeks

Sphenoid bone
Brain
Posterior lobe of the pituitary gland
Anterior lobe of the pituitary gland
Secondary palate
Tongue

Adult

Hypothalamus
Sphenoid bone
Pituitary gland

Figure 37.2
Development of the adrenal glands

Week 5

GI tract
Sympathetic ganglia (developing)
Abdominal mesenchyme
Foetal cortex of the suprarenal gland

Week 6

Neural crest cells migrate to suprarenal cortex from sympathetic ganglion

Week 8

Neural crest cells penetrate suprarenal cortex and form the suprarenal gland's medulla
Kidney

Foetal period

Mesenchymal cells surround the foetal cortex and form the permanent cortex
Kidney

One year after birth

Vertebrae
Permanent cortex differentiates to form the layers of zona glomerulosa, zona fasciculata and zona reticularis
Kidney

Time period: day 24 to birth

Introduction

The glands of the endocrine system begin to form during the embryonic period and continue to mature during the foetal period.

Functional development can be detected by the presence of the various hormones in the foetal blood, generally in the second trimester of pregnancy.

The development of the gonads, pancreas, kidneys and placenta are covered elsewhere in this book.

Embryology at a Glance, First Edition. Samuel Webster and Rhiannon de Wreede.

Pituitary gland

Also known as the **hypophysis**, the pituitary gland develops from two sources. An outpocketing of oral ectoderm appears in week 3 in front of the buccopharyngeal membrane (Figure 37.1). This forms the **hypophysial diverticulum** (or Rathke's pouch), which will become the anterior lobe.

The second source is an extension of neuroectoderm from the diencephalon, called the **neurohypophysial diverticulum** (or infundibulum). The infundibulum grows downwards, developing into the posterior lobe. These two parts grow towards one another and by the second month the hypophysial diverticulum is isolated from its ectodermal origin and lies close to the infundibulum.

Growth hormone secreted by the pituitary gland can be detected from 10 weeks.

Hypothalamus

The hypothalamus begins to form in the walls of the diencephalon (see Chapter 42), with nuclei developing here that will be involved in endocrine activities and homeostasis.

Pineal body

The pineal body first appears as a diverticulum in the caudal part of the roof of the diencephalon. It becomes a solid organ as the cells here proliferate.

Adrenal glands

The adrenal (or suprarenal) glands develop from two cell types. The cells of the **cortex** differentiate from mesoderm of the posterior abdominal wall near the site of the developing gonad (Figure 37.2). The adrenaline and noradrenaline secreting cells of the **medulla** are derived from migrating neural crest cells that formed a sympathetic ganglion nearby. These cells become surrounded by the cell mass of the cortex.

The foetal cortex produces a steroid precursor of oestrogen that is converted to oestrogen by the placenta. More mesenchymal cells surround the foetal cortex and will become the layers of the permanent cortex.

The adrenal glands are exceptionally large in the foetus because of the size of the cortex which regresses after birth. Substances secreted from the adrenal glands are involved in the maturation of other systems of the embryo, such as the lungs and reproductive organs.

Thyroid gland

This is the first endocrine gland to develop, beginning at about 24 days between the first and second pharyngeal pouches from a proliferation of endodermal cells of the gut tube. It begins as a hollow thickening of the midline where the future tongue will develop. It eventually becomes solid and then splits into its two lobes.

As the thyroid descends into the neck it remains connected to the tongue via the **thyroglossal duct** with an opening on the tongue called the **foramen cecum**. The duct degenerates between weeks 7 and 10 and the thyroid reaches its end location anterior to the trachea by week 7. If parts of the duct remain the person may also have a **pyramidal lobe**. This is quite common and seen in about 50% of the population.

C cells (or parafollicular cells) are derived from neural crest cells that invade the **ultimobranchial body** (a fifth pharyngeal pouch derivative; see Chapter 41).

Parathyroid glands

The inferior parathyroid glands develop from epithelium (endoderm) of the dorsal wing of the **third pharyngeal pouch**. The cells here move with the migration of the thymus gland into the neck (see Chapter 40). When this connection breaks down they become located on the dorsal surface of the thyroid gland.

Endoderm cells of the dorsal wing of the **fourth pharyngeal arch** begin to collect and differentiate to form the superior parathyroid glands (initially the superior parathyroid glands are inferior to the inferior parathyroid glands). These cells are associated with the developing thyroid gland and migrate with it, but for a shorter distance than the cells of the inferior parathyroid glands (see Chapter 41). They also rest on the dorsal surface of the thyroid, but generally more medially and posteriorly.

Clinical relevance
Pituitary gland

Congenital hypopituitarism is a decrease in the amount of one or more of the hormones secreted by the pituitary gland. Symptoms are wide ranging, depending upon which hormones are affected. The cause is often hypoplasia of the gland or complications with delivery. Treatment is commonly oral or injection replacement of the insufficient hormones.

Adrenal glands

Congenital adrenal hyperplasia is an autosomal recessive disease causing excessive production of steroids, with 95% of patients deficient in the enzyme **21-hydroxylase** (required in the production of adrenal secretions). There are degrees of severity and this can cause ambiguous genitalia and infertility. Various treatment options are available and can include glucocorticoids, sex hormone replacement and genital reconstructive surgery.

Thyroid gland

Congenital hypothyroidism is a deficiency in thyroid hormone production. Symptoms include excessive sleeping and poor feeding. Newborn infants are screened for this and if this deficiency is found treatment is a daily thyroxine tablet.

Ectopic thyroid tissue left behind during migration is relatively common but asymptomatic. Parts of the thyroglossal duct may persist and form a midline, moveable cyst in a child.

Parathyroid glands

Hypoparathyroidism is an absence of parathyroid hormone. Symptoms are wide ranging but often not diagnosed until 2 years of age. They include seizures and poor growth. Treatment includes vitamin D and calcium supplements.

Ectopic parathyroid tissue left behind during migration is relatively common but asymptomatic. It is more common for the inferior parathyroid glands.

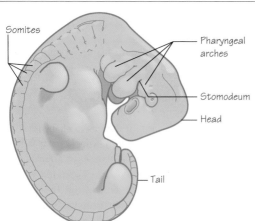

Figure 38.1
Lateral aspect of the 4 week embryo with the pharyngeal arches visible lying cranial to the somites

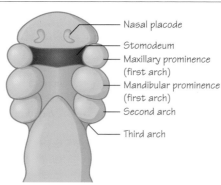

Figure 38.2
Ventral aspect of the cranial part of the embryo showing the structures developing around the stomodeum, including the first, second and third pharyngeal arches

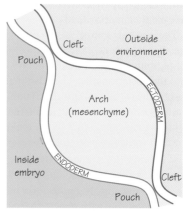

Figure 38.3
An outline of the relationship between each pharyngeal arch and each pharyngeal cleft and pouch

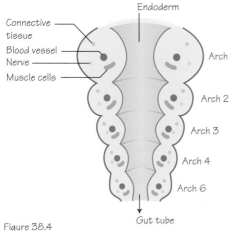

Figure 38.4
The arches appear and develop at different rates, so the first arches are more developed by the time the sixth arches appear. Each arch has its own nerve, artery, connective tissue cells and muscle cells

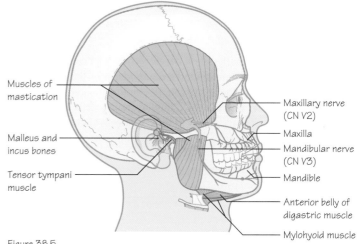

Figure 38.5
Structures derived from the cells of the first pharyngeal arch

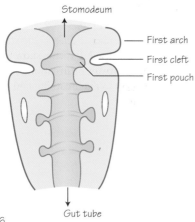

Figure 38.6
Week 6. The clefts between most of the pharyngeal arches have disappeared, but the first cleft remains as the external acoustic meatus. The first pouch will form the pharyngotympanic tube

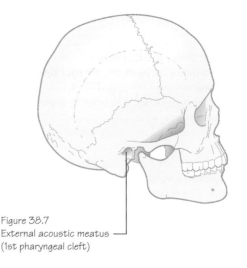

Figure 38.7
External acoustic meatus (1st pharyngeal cleft)

Embryology at a Glance, First Edition. Samuel Webster and Rhiannon de Wreede.

Time period: day 21 onwards

Introduction

Pharyngeal (or branchial) **arches** are paired structures that develop in the ventrolateral parts of the head of the embryo (Figures 38.1 and 38.2). Six arches will form and contribute to the development of head and neck structures, although arch V is ignored as it fails to appear in human embryos. In this chapter we concentrate on arch I and its derivatives.

Each pharyngeal arch is a bud, or bar of mesenchymal tissue, with clefts separating the arches externally, and pouches separating them internally (Figure 38.3). Pharyngeal pouches develop internally as blebs of the foregut at the level of the pharynx.

Each pharyngeal arch consists of mesenchyme from paraxial and lateral plate mesoderm and receives an influx of neural crest cells. Neural crest cells from rhombomeres 1 and 2 (see Chapter 43 and Figure 43.4) migrate into the first pharyngeal arch. Hox genes, important in the organisation of the segmentation of vertebrates and in setting up the anteroposterior axis, are also important in neural crest cell migration here.

Each arch has its own nerve, artery, connective tissue cells and muscle cells (Figure 38.4).

Arch I

In week 4 a depression in the surface ectoderm of the embryo forms in the future face, the **stomodeum** (Figure 38.2). It is continuous with the gut tube and will become the mouth. It forms the centre of the face early in development, and surrounding it are the first pair of pharyngeal arches.

The first arch can be divided into a dorsal **maxillary** process and a ventral **mandibular** process (Figure 38.2). The mandibular process contains Meckel's cartilage, which provides the horseshoe-shaped model for the mandible to form around and later degenerates after contributing to the connective tissue structures of the first arch, such as the incus and malleus of the middle ear (see box). The mandible forms by intramembranous ossification, rather than endochondral ossification.

Structures formed from the first pharyngeal arch (Figure 38.5)

Bones	Incus, malleus, maxilla, mandible (also squamous part of the temporal bone, zygomatic bone, palatine bones)
Ligaments	Sphenomandibular ligament, anterior ligament of malleus
Muscles	Muscles of mastication (temporalis, masseter, pterygoids), anterior belly of digastric, mylohyoid, tensor veli palatini and tensor tympani
Nerve	Trigeminal nerve (CN V) (maxillary and mandibular branches – V_2 and V_3)
Blood supply	Maxillary artery (first branch from the first aortic arch)

Note how the mandibular branch of the trigeminal nerve (CN V) supplies motor fibres to the muscles of mastication in the adult. All these structures are derived from first arch cells. The trigeminal nerve is the cranial nerve of the first arch and is the major sensory nerve of the skin of the face. Only its maxillary and mandibular branches (V_2 and V_3) supply structures derived from the first arch, however.

Ectoderm and endoderm from this arch also form the mucous membrane and glands on the anterior two-thirds of the tongue.

Cleft I

The only cleft of any embryological importance in humans is present in week 5 and will form the **external auditory meatus** of the ear (Figures 38.6 and 38.7). Considering that the overlap between arch I and II structures is at the ear (note the origins of the malleus, incus and stapes bones) this makes sense. The other clefts largely disappear.

The pharyngeal cleft has an ectodermal surface (Figure 38.3).

Pouch I

The first pharyngeal pouch becomes a lengthy ingrowth which becomes the **tubotympanic recess** (Figure 38.6). With other pharyngeal structures this will form the pharyngotympanic tube. The first pouch extends towards the first cleft. The tubotympanic recess can be divided into dorsal and ventral parts, forming the middle ear cavity and auditory tube, respectively.

The pharyngeal pouch has an endodermal surface (Figure 38.3).

Clinical relevance

Facial abnormalities can be caused by failures of neural crest cell migration and are associated with other abnormalities including cardiac defects. Failure of neural crest cells to migrate into the arches can affect any of the structures that arise from pharyngeal arch I.

Treacher Collins syndrome is characterised by craniofacial abnormalities, including a cleft palate, a small mandible and malformed or absent ears. Linked to a mutation on chromosome 5 it can be inherited but can also arise from a random mutation. This mutation prevents neural crest cell migration into the first pharyngeal arch. Deformities can vary in severity and surgery is often required. Hearing problems are common, as you might imagine given the ear structures formed by the first arch.

Robin sequence is also characterised by facial abnormalities but is a sequence of events rather than a genetic abnormality, and its cause is not completely understood. Patients show a small mandible, cleft palate and upper airway obstruction. Nasopharyngeal cannulation is often necessary to aid breathing and feeding. Surgical treatment is also required.

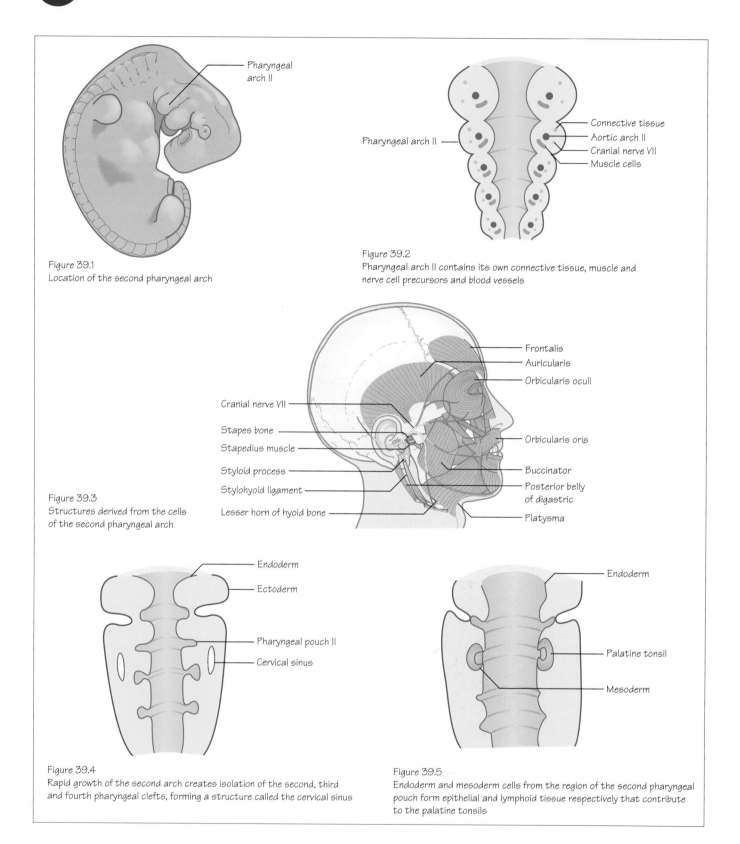

Figure 39.1
Location of the second pharyngeal arch

Pharyngeal
arch II

Figure 39.2
Pharyngeal arch II contains its own connective tissue, muscle and
nerve cell precursors and blood vessels

Pharyngeal arch II

Connective tissue
Aortic arch II
Cranial nerve VII
Muscle cells

Figure 39.3
Structures derived from the cells
of the second pharyngeal arch

Cranial nerve VII
Stapes bone
Stapedius muscle
Styloid process
Stylohyoid ligament
Lesser horn of hyoid bone

Frontalis
Auricularis
Orbicularis oculi
Orbicularis oris
Buccinator
Posterior belly
of digastric
Platysma

Figure 39.4
Rapid growth of the second arch creates isolation of the second, third
and fourth pharyngeal clefts, forming a structure called the cervical sinus

Endoderm
Ectoderm
Pharyngeal pouch II
Cervical sinus

Figure 39.5
Endoderm and mesoderm cells from the region of the second pharyngeal
pouch form epithelial and lymphoid tissue respectively that contribute
to the palatine tonsils

Endoderm
Palatine tonsil
Mesoderm

Embryology at a Glance, First Edition. Samuel Webster and Rhiannon de Wreede.

Time period: day 21 onwards

Introduction

The second arch forms caudally to the first arch (Figure 39.1). Pharyngeal arches I and II are bigger than III and IV. Arch II grows rapidly and inferiorly to cover the smaller arches forming the **operculum**. This growth forms a 'lid' over the other arches and creates the smooth covering of the neck.

Arch II

Highlighting the overlap between arches I and II at the ear, the stapes bone is formed from the connective tissue element of the second arch, whereas the malleus and incus bones develop from the first arch. Likewise, the tensor tympani muscle of the ear forms from the first arch but the stapedius muscle is derived from the second arch.

The second arch also contributes to the bony styloid process of the skull, cranial parts and lesser horn (cornu) of the hyoid bone and the stylohyoid ligament that connects them. The remainder of the hyoid bone develops from the third arch, highlighting another overlap between arches.

The cranial nerve of the second arch is the facial nerve (CN VII; Figure 39.2). Consider this when looking at the anatomical route that CN VII neurons take in relation to the middle and inner ear and the styloid process. The facial nerve will innervate the muscle and mucosal derivatives of the second arch (e.g. the mucosa of the tongue).

The facial nerve is also the nerve to the muscles of facial expression, and these muscles are derived from the muscle block of the second arch.

The blood vessels of the pharyngeal arches are described in the cardiovascular chapters, and in general the second aortic arches are lost. The second aortic arch forms a stapedial artery that links internal and external carotid arteries, but normally does not persist beyond foetal life. The stapedial artery passes through stapes, forming its foramen. Of note, the middle meningeal artery is associated with the development of the first and second aortic arches (see box).

Cleft II

The second pharyngeal cleft becomes isolated from the external environment by growth of the second arch when it forms the operculum. Consequently, it forms a sinus with the third and fourth clefts, the **cervical sinus,** lined with ectodermal epithelia (Figure 39.4).

With further growth this sinus normally disappears entirely. The ectoderm of the sinus is eventually involved in the epithelial cells of Hassall's corpuscles found in the thymus gland.

Pouch II

The endoderm of the second pharyngeal pouch proliferates and pushes into the mesenchyme beneath to form lymphoid tissue. The palatine tonsils form as a result, with the lymphoid tissue derived from the mesoderm, and epithelial cell-lined crypts (Figure 39.5).

Clinical relevance

A persistent stapedial artery presents as a pulsing in the ear or accidentally found during surgery, but can cause hearing loss.

Pharyngeal cleft (or branchial) cysts can be found after birth, in which cysts are located under the platysma muscle, laterally, and anterior to the sternocleidomastoid muscle. The cysts commonly enlarge slowly and appear much later in life. The cysts have formed from the remnants of the pharyngeal clefts that normally combine to briefly create the cervical sinus. The cyst may open externally as a sinus.

Congenital facial paralysis is rare but results in a lack of facial expression and can affect lateral eye movement. **Goldenhar syndrome** affects both first and second pharyngeal arches and the affected individual can show facial palsy, but also malformations of other facial bones including the maxilla and zygomatic bones.

Structures formed from the second pharyngeal arch (Figure 39.3)

Bones	Stapes, styloid process of temporal bone, lesser horn and superior part of the body of the hyoid bone
Muscles	Muscles of facial expression, stapedius, mylohyoid, posterior belly of digastric, auricular, buccinator, platysma
Ligaments	Stylohyoid ligament
Nerve	Facial nerve (CN VII)
Blood supply	Hyoid artery (foetal), stapedial artery (foetal)

40 Head and neck: arch III

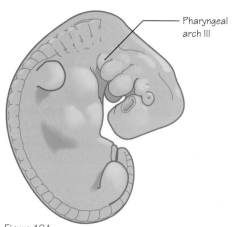

Figure 40.1
Location of the third pharyngeal arch

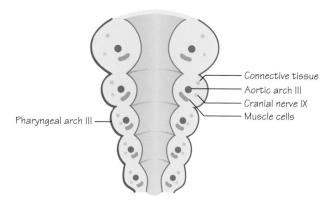

Figure 40.2
Pharyngeal arch III contains its own connective tissue, muscle and nerve cell precursors and blood vessels

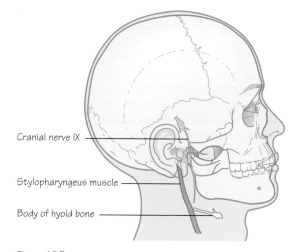

Cranial nerve IX

Stylopharyngeus muscle

Body of hyoid bone

Figure 40.3
Structures derived from cells of the third pharyngeal arch

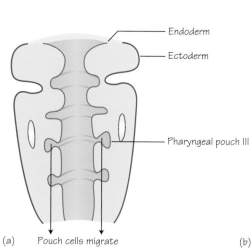

Endoderm
Ectoderm

Pharyngeal pouch III

(a) Pouch cells migrate

3rd pouch cells migrate to posterior thyroid gland to become inferior parathyroid glands and thymus

Thyroid gland
Parathyroid glands (behind thyroid gland)
Thymus

(b)

Figure 40.4
Cells of the third pharyngeal pouch form the thymus and the inferior parathyroid glands

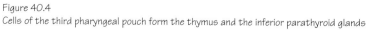

Embryology at a Glance, First Edition. Samuel Webster and Rhiannon de Wreede.

Time period: day 28 onwards

Introduction

The third and fourth pharyngeal arches form in the fifth week (Figure 40.1). Both are considerably smaller than the first two arches and have fewer derivatives.

Arch III

The connective tissue element of the third arch will become the greater horn (cornu) and caudal parts of the hyoid bone (Figures 40.2 and 40.3). The second arch formed the other parts of the hyoid (see Chapter 39).

The muscle element becomes the stylopharyngeus muscle, passing from the styloid process to the pharyngeal constrictor muscles. The third arch is involved in the development of pharyngeal structures, and its nerve is the glossopharyngeal nerve (CN IX). Motor fibres to the stylopharyngeus muscle are supplied by CN IX, and this nerve also supplies motor fibres and sensory fibres to the pharyngeal plexus for innervation of the muscles and mucosa of the pharynx.

The pharyngeal plexus forms from glossopharyngeal and vagus nerve (CN X) fibres. The vagus nerve is the nerve of the fourth and sixth arches, and the overlap between structures of the head and neck derived from different pharyngeal arches is apparent in adult anatomy.

The third aortic arches, the arteries of the third pharyngeal arches (one on either side of the neck), form major parts of the adult common and internal carotid arteries (see Figure 41.3). The internal carotid arteries are formed from the third aortic arch on either side in combination with the dorsal aortae (see box).

Structures formed from the third pharyngeal arch (Figure 40.3)

Bones	Greater horn and the inferior part of the body of the hyoid bone
Muscles	Stylopharyngeus muscle
Nerve	Glossopharyngeal nerve (CN IX)
Blood supply	Common carotid artery and the first part of the internal carotid artery

---

Cleft III

The third pharyngeal cleft is incorporated into the cervical sinus and eventually obliterated (see Chapter 39).

Pouch III

The endoderm lining the third pharyngeal pouch is involved in the development of endocrine glands in the neck. The third pouch can be split into a ventral wing and a dorsal wing.

Cells of the ventral parts of the left and right third pharyngeal arches come together and migrate caudally to form the majority of the **thymus gland** (Figure 40.4). Neural crest cells are also involved in development of the thymus gland, which continues to mature after birth.

The cells of the dorsal part of the third pouch differentiate into parathyroid gland cells in the sixth week, and migrate with the cells of the thymus gland to descend into the neck (Figure 40.4). These cells will form the **inferior parathyroid gland** and later separate from the thymus gland, moving dorsally to it as found in the adult.

Clinical relevance

Defects in pharyngeal arch III related development affect the formation of the thymus and parathyroid glands. Neural crest cell migration or proliferation problems may affect thymus development.

Congenital hypoparathyroidism is a condition in which parathyroid hormone (PTH) is secreted in low levels, potentially caused by a failure of one or more parathyroid glands to develop normally. Symptoms are wide ranging but often not diagnosed until 2 years of age, and include seizures and poor growth. Treatment options include administering vitamin D and calcium supplements.

Ectopic parathyroid tissue left behind during migration of the cells of the third pouch is relatively common but asymptomatic. This may be more common for the inferior parathyroid gland.

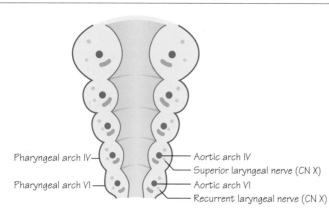

Figure 41.1
Pharyngeal arches IV and VI contain their own connective tissue, muscle and nerve cell precursors and blood vessels

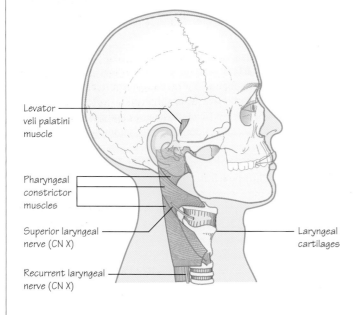

Figure 41.2
Structures derived from the cells of the fourth and sixth pharyngeal arches

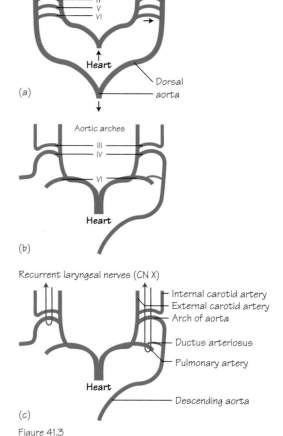

Figure 41.3
The aortic arches discussed in Chapter 27 are the arteries of the pharyngeal arches.
(a) The aortic arches bridge the outflow from the heart with the paired dorsal aortae at this stage
(b) With development some aortic arches are lost, and some combine
(c) The recurrent laryngeal nerves of the sixth arch loop around the ductus arteriosus (and aorta) on the left and around the subclavian artery on the right

Embryology at a Glance, First Edition. Samuel Webster and Rhiannon de Wreede.

Time period: day 28 onwards

Introduction

The fourth and sixth pharyngeal arches are often discussed together (Figure 41.1). Although separate, they are linked by the structures that they will form in the neck. A fifth pharyngeal arch forms briefly but it quickly degenerates leaving no remnants in humans. The fourth and sixth arches eventually fuse.

Neural crest cells from rhombomere 6 (see Chapter 43) migrate into the fourth pharyngeal arch (as well as the third arch, as we saw in Chapter 40).

The connective tissue elements of the fourth and sixth arches combine and fuse to form all the laryngeal cartilages, except for the epiglottis (Figure 41.2). Not surprisingly, the muscle cells of the fourth and sixth arches form the muscles of the larynx and pharynx, including the intrinsic muscles of the larynx (fourth arch), the constrictors of the pharynx (sixth arch) and one of the muscles of the palate (levator veli palatini, sixth arch).

From your anatomical knowledge of the pharynx and larynx you should already be guessing that the cranial nerve associated with both of these pharyngeal arches is the **vagus nerve** (CN X).

The fourth arch is associated with the **superior laryngeal branch** of the vagus, and the sixth arch is associated with the **recurrent laryngeal nerve**. The muscles of the sixth arch are the intrinsic muscles of the larynx, and these muscles receive motor innervation from the recurrent laryngeal nerve. The vagus nerve also contributes to the pharyngeal plexus with the glossopharyngeal nerve, innervating pharyngeal mucosa and musculature.

You may also recall the lengthy detour that the recurrent laryngeal nerve takes in the adult, descending from the neck to loop around the subclavian artery on the right and the arch of the aorta on the left, before ascending between the oesophagus and trachea to the larynx. The right subclavian artery forms from the right fourth aortic arch (this is the artery of the fourth pharyngeal arch) and the arch of the adult aorta forms from the left fourth aortic arch (Figure 41.3). The left sixth aortic arch forms the left pulmonary artery and its connection to the aorta: the ductus arteriosus. The right sixth aortic arch forms the right pulmonary artery but its link to the embryonic right dorsal aorta is lost.

The nerves form here too, at the same levels. Evidence that the recurrent laryngeal nerve is the nerve of the sixth arch can be found in the adult, as the left recurrent laryngeal nerve passes around the ligamentum arteriosum, the remnant of the ductus arteriosus, the sixth arch artery linking the left pulmonary artery with the aorta (Figure 41.3). The nerves are prevented from ascending fully into the neck as the embryo grows and lengthens by the ductus arteriosus on the right and the subclavian artery on the left (see box).

Structures formed from the fourth and sixth pharyngeal arches (Figure 41.2)

Cartilage	Thyroid, cricoid, arytenoid, corniculate, cuneiform
Muscles	Arch IV: cricothyroid, levator veli palatini and the pharyngeal constrictors
	Arch VI: intrinsic muscles of the larynx (except cricothyroid)
Nerve	Arch IV: superior laryngeal, branch of the vagus nerve
	Arch VI: recurrent laryngeal, branch of the vagus nerve
	At this level in the embryo we see differences in the cardiovascular system develop between left and right sides.
Blood	Arch IV: left, aortic arch; right, subclavian artery
	Arch VI: left, pulmonary artery and ductus arteriosus; right, pulmonary artery

---

Cleft IV

The fourth pharyngeal cleft is incorporated into the cervical sinus with the third cleft and eventually obliterated (see Figure 39.4).

Pouch IV

From the endoderm of the fourth pharyngeal pouch the **superior parathyroid glands** form. From the fifth pouch (often considered part of the fourth pouch) the **ultimobranchial body** (or ultimopharyngeal body) forms. Cells of the ultimobranchial body invade the thyroid gland and differentiate into the C cells (or parafollicular cells) which will produce calcitonin (Figure 41.3).

There is no pharyngeal pouch VI.

Clinical relevance

Congenital cricoid cartilage abnormalities tend to affect the size or shape of the cartilage. This can lead to **congenital subglottic stenosis**, causing difficulty in breathing, and requires surgery.

Laryngomalacia is a common congenital laryngeal abnormality and patients have a larynx that collapses during breathing causing significant breathing difficulties. Other symptoms include a noise that can be heard during inspiration and gastroesophageal reflux. In most cases as the larynx continues to develop the symptoms are eased and are insignificant by 2 years of age. In severe cases surgery may be required.

Abnormal development of pharyngeal arch IV can affect the parathyroid glands and subsequently the quantity of hormones that these cells produce. Low levels of parathyroid hormone (**hypoparathyroidism**) can result in hypocalcaemia (low serum calcium levels). DiGeorge syndrome is a known cause of hypoparathyroidism. Symptoms are wide ranging and include muscle cramps, pain in the face and abdomen, dry hair, nails and skin and weak tooth enamel. Treatment includes calcium, vitamin D and synthetic parathyroid hormone supplements.

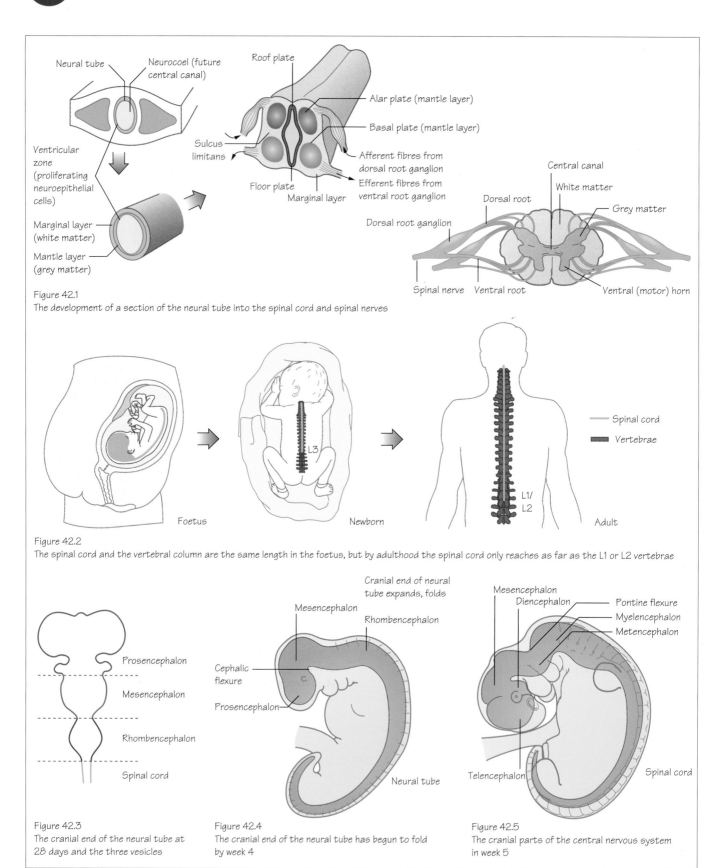

Figure 42.1
The development of a section of the neural tube into the spinal cord and spinal nerves

Figure 42.2
The spinal cord and the vertebral column are the same length in the foetus, but by adulthood the spinal cord only reaches as far as the L1 or L2 vertebrae

Figure 42.3
The cranial end of the neural tube at 28 days and the three vesicles

Figure 42.4
The cranial end of the neural tube has begun to fold by week 4

Figure 42.5
The cranial parts of the central nervous system in week 5

Embryology at a Glance, First Edition. Samuel Webster and Rhiannon de Wreede.

Time period: day 22 to postnatal development

Introduction

Ectoderm is induced by the notochord to form neuroectoderm during neurulation (see Chapter 15). This neuroectoderm in turn produces the neural tube and neural crest cells from which the central nervous system develops. The central nervous system comprises the brain and spinal cord.

Spinal cord

The caudal end of the neural tube continues to elongate and form the spinal cord. A lumen through the centre of the spinal cord, the neurocoel (or neural canal), forms by week 9 and will become the central canal. The neurocoel is lined with thickening layers of **neuroepithelia** known as the **ventricular zone** (Figure 42.1) or ependymal layer.

Cells of the ventricular zone differentiate into neuroblasts and glioblasts (or spongioblasts). Glioblasts will form supporting cells, or neuroglial cells. Neuroblasts become neurons and migrate out to form the **mantle** (or intermediate) **layer** of cell bodies, which will form the **grey matter**. Surrounding this layer is the outer **marginal layer** that carries the axons from the neurons in the mantle layer, and will become **white matter** (Figure 42.1).

Dorsal root ganglia (also known as spinal ganglia) outside the neural tube formed from neural crest cells extend central processes which grow into the neural tube. Some form synapses with neurons in the mantle layer (grey matter) while other central processes ascend within the marginal layer (white matter).

Further distinction of the mantle layer occurs as the ventral area becomes the **basal plates** and the dorsal part becomes the **alar plates**, forming the ventral motor horns and dorsal sensory horns, respectively (Figure 42.1). They are divided by a groove, the **sulcus limitans**. Between these horns the **intermediate horn** develops containing neurons of the sympathetic nervous system between spinal levels T1 and L2 (or L3).

Spinal nerves form as the axons of dorsal and ventral roots from the spinal cord combine, with input from peripheral processes of the dorsal root ganglia and, where level appropriate, input from the autonomic nervous system.

In the embryo the spinal cord reaches to the caudal end of the vertebral column. Towards the end of the embryonic period the coccygeal vertebrae are reduced and the vertebral column grows rapidly causing the tail of the spinal cord to extend to the level of the L3 vertebra by birth, and by adulthood commonly only reaches to between L1 and L2 (Figure 42.2).

Brain

Three vesicles develop in the cranial end of the neural tube (Figure 42.3). The distinct alar and basal plates we saw in the spinal cord are retained as the hindbrain and midbrain develop, whereas in the forebrain the dorsal alar plates expand and the ventral basal plates degenerate.

The vesicles become the **prosencephalon** (forebrain), **mesencephalon** (midbrain) and **rhombencephalon** (hindbrain). As the vesicles grow the tube folds (Figure 42.4). In the fifth week the three vesicles become five, with the forebrain and hindbrain both splitting into two, forming the **telecephalon** and **diencephalon** from the forebrain, and from the hindbrain the **metencephalon** and the **myelencephalon** form (Figure 42.5).

The dorsal telencephalon of the forebrain will grow rapidly to form the **cerebral cortex**, and the ventral telencephalon will become the **basal ganglia**.

The diencephalon will form the **optic cup** and **stalk, pituitary gland, thalamus, hypothalamus** and **pineal body**.

The caudal part of the hindbrain, the myelencephalon, becomes the **medulla oblongata**. The metecephalon develops into the **pons** ventrally and the **cerebellum** dorsally. The midbrain, together with the pons and medulla oblongata form the **brainstem**.

Ventricles of the brain form from the lumen of the neural tube. Ventricles contain cerebrospinal fluid produced by **choroid plexuses**. **Lateral ventricles** of the telencephalon link with the **third ventricle** of the diencephalon, which is connected to the **fourth ventricle** of the myelencephalon through the cerebral aqueduct of the mesencephalon. The fourth ventricle is continuous with the central canal of the spinal cord.

Neural crest cells

Neural crest cells migrate to both sides of the spinal cord and form the dorsal root ganglia (sensory ganglia) of the spinal nerves. In the hindbrain area they contribute to cranial nerve ganglia, sensory ganglia of CN V, VII, VIII, IX and X.

Meninges

Of the connective tissue layers that surround the brain, the **pia mater** and **arachnoid mater** are mesoderm and **neural crest cell** derived, whereas the **dura mater** is only mesoderm derived.

Clinical relevance

Encephalocoele is the herniation of dura, potentially containing brain tissue, through a midline skull defect, caused by incomplete closure of the neural tube. Repaired surgically, recovery depends upon whether or not neural tissue was enclosed in the encephalocoele.

Anencephaly describes incomplete closure at the cranial end of the neural tube, resulting in a complete failure of forebrain formation. This is lethal and most infants with this condition are stillborn.

Hydrocephalus is a condition resulting from excess cerebrospinal fluid (CSF) resulting from a blockage of flow, failure to reabsorb or increased production. Conditions include increasing head circumference and vomiting, but ventricular dilation can be identified before symptoms become visible. Treatment options include inserting a shunt to open a blockage or to redirect the CSF to an area where it can be reabsorbed.

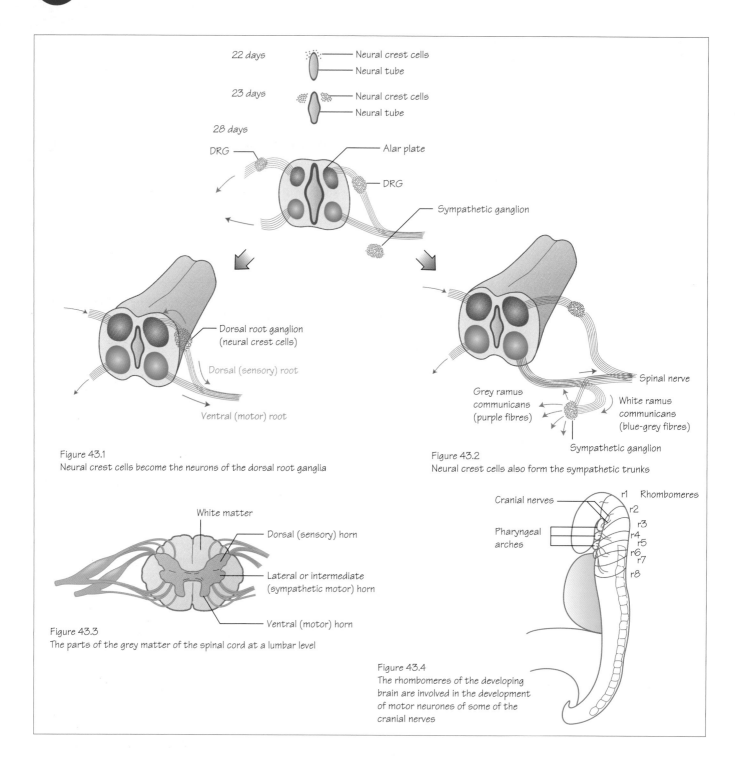

22 days — Neural crest cells / Neural tube

23 days — Neural crest cells / Neural tube

28 days

DRG — Alar plate — DRG — Sympathetic ganglion

Dorsal root ganglion (neural crest cells)

Dorsal (sensory) root

Ventral (motor) root

Figure 43.1
Neural crest cells become the neurons of the dorsal root ganglia

Spinal nerve

Grey ramus communicans (purple fibres)

White ramus communicans (blue-grey fibres)

Sympathetic ganglion

Figure 43.2
Neural crest cells also form the sympathetic trunks

White matter

Dorsal (sensory) horn

Lateral or intermediate (sympathetic motor) horn

Ventral (motor) horn

Figure 43.3
The parts of the grey matter of the spinal cord at a lumbar level

Cranial nerves

Pharyngeal arches

r1 Rhombomeres
r2
r3
r4
r5
r6
r7
r8

Figure 43.4
The rhombomeres of the developing brain are involved in the development of motor neurones of some of the cranial nerves

Embryology at a Glance, First Edition. Samuel Webster and Rhiannon de Wreede.

Time period: day 27 to birth

Introduction

The peripheral nervous system develops in tandem with the brain and spinal cord. It connects the central nervous system to structures of the body as they form and includes the spinal nerves, cranial nerves and autonomic nervous system.

This process begins with neurulation (see Chapter 15), when ectoderm is induced by the notochord to form neuroectoderm. This neuroectoderm in turn produces neuroblasts (primitive neurons) and neural crest cells.

Spinal nerves

Neural crest cells migrate out from the neural tube, passing towards multiple targets throughout the embryo (see Chapter 16). Some neural crest cells only migrate a little way from the developing spinal cord, collect together and differentiate to form neurons of the **dorsal root ganglia** (Figure 43.1). Located bilaterally to the spinal cord, the dorsal root ganglia send afferent processes back towards the alar plate of the spinal cord (see Figure 42.1), eventually passing to the dorsal horn. The dorsal root ganglia also send processes out to run alongside processes of neurons of the ventral root. Their combined bundle of neuronal axons become the spinal nerve.

Dorsal root ganglia contain the cells bodies of sensory neurons from afferent spinal nerve fibres. It could be considered that these ganglia are grey matter of the spinal cord that have moved out to the peripheral nervous system.

Neurons in the (ventral) basal plates of the developing spinal cord send fibres outwards from the cord to meet the fibres from the dorsal root ganglion. Fibres grouping to leave the spinal cord form the ventral horn, the motor fibres of the spinal nerves (Figures 42.1 and 43.1).

The mixed collection of motor and sensory fibres split again almost as soon as they meet, forming two bundles of fibres: the **dorsal ramus** and the **ventral ramus**. The dorsal ramus passes dorsally to the deep axial musculature of the back, the synovial vertebral joints and the skin of this region. The ventral ramus sends fibres to the ventral and lateral parts of the trunk and to the upper and lower limbs, depending upon the spinal level.

Neural crest cells migrate to the new axons that extend away from the central nervous system and wrap themselves around them, differentiating to become **neurolemmal (Schwann) cells**. Myelin within these cells causes the fibres to appear white. This is apparent from around week 20.

Dermatomes

Spinal nerves develop at the level of each somite, and the segmented organisation of embryonic development can be seen in the adult segmented pattern of cutaneous dermatomes (see Chapter 20).

A dermatome is a region of skin that is predominantly supplied by the sensory component of one spinal nerve. The dermatomes are named according to the spinal nerve that supplies them (see Figure 20.5).

Autonomic nervous system

Some neural crest cells migrating out from the neural tube collect dorsolaterally to the dorsal aorta in the thorax. They differentiate and become chains of sympathetic ganglia connected by longitudinal nerve fibres running cranially and caudally (Figure 43.2). Neuroblasts migrate from the ganglia cranially into the neck and caudally into the abdomen and pelvis to complete the **sympathetic trunks**.

Other neural crest cells migrate ventrally to the aorta to form the **preaortic ganglia** such as the coeliac and mesenteric ganglia, while others migrate towards organs such as the lungs, heart and gastrointestinal tract to form sympathetic organ plexuses.

Sympathetic neurons of the developing spinal cord in the intermediolateral cell column (intermediate or lateral horn) of thoracolumbar segments T1–L2 (Figure 43.3) send axons out from the spinal cord through the ventral root to each trunk ganglion. These axons form the **white ramus communicans** passing between the spinal nerve and the sympathetic ganglion (Figure 43.2). These preganglionic sympathetic fibres either synapse with neurons in the ganglion, ascend or descend to synapse in a ganglion of a different level, or pass through the ganglion/ganglia without synapsing to run towards preaortic (or prevertebral) ganglia or organs as splanchnic nerves.

Postganglionic fibres pass onwards to viscera or group together to pass back to a spinal nerve as a **grey ramus communicans** (unmyelinated). Grey ramus communicans are found at all spinal levels and allow postganglionic fibres to pass with other nerves to reach structures, for example in the limbs.

Neurons of the parasympathetic nervous system form in the brainstem and sacral part of the spinal cord. The parasympathetic nuclei in the brainstem contribute preganglionic fibres to the oculomotor (CN III), facial (CN VII), glossopharyngeal (CN IX) and vagus (CN X) nerves. Preganglionic parasympathetic fibres of the sacral spinal cord form pelvic splanchnic nerves.

Postganglionic parasympathetic neurons differentiate from neural crest cells and can be found either in ganglia or near the viscera.

Cranial nerves

Cranial nerves develop in a similar way to the spinal nerves, with motor nuclei differentiating from the neuroepithelium and sensory nuclei forming outside the brain.

The olfactory nerve (CN I) and the optic nerve (CN II) are linked with the telencephalon and diencephalon, respectively. The olfactory nerve connects to the olfactory bulb, a growth from the prosencephalon. The oculomotor nerve (CN III) originates from the dorsal midbrain (mesencephalon) and the trochlear nerve (CN IV) from the ventral midbrain (metencephalon). Similarly, CN V–VIII arise from the metencephalon and CN IX–XII arise from the myelencephalon.

Cranial nerves IV–VII and IX–XII develop from the hindbrain which has been divided into eight sections called **rhombomeres** by day 25 (Figure 43.4). Rhombomeres appear ventral to the cephalic flexure by day 29 and will form the motor neurons for these cranial nerves (note that these are the cranial nerves with motor function, hence the skipping of CN VIII).

Neural crest cells form the parasympathetic neurons for CN III, VII, IX and X.

The development and structure of the cranial nerves is very similar to the spinal nerves, but remember that not all cranial nerves carry both sensory and motor neurons.

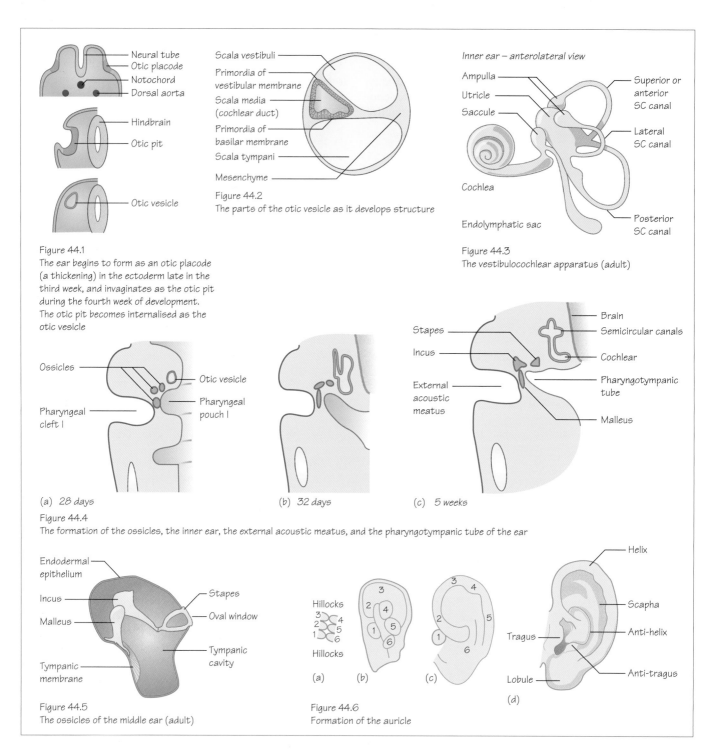

Figure 44.1
The ear begins to form as an otic placode (a thickening) in the ectoderm late in the third week, and invaginates as the otic pit during the fourth week of development. The otic pit becomes internalised as the otic vesicle

Figure 44.2
The parts of the otic vesicle as it develops structure

Figure 44.3
The vestibulocochlear apparatus (adult)

(a) 28 days   (b) 32 days   (c) 5 weeks

Figure 44.4
The formation of the ossicles, the inner ear, the external acoustic meatus, and the pharyngotympanic tube of the ear

Figure 44.5
The ossicles of the middle ear (adult)

Figure 44.6
Formation of the auricle

Time period: 22 day to birth

Internal ear

The function of the internal ear is to receive sound waves and interpret them into nerve signals, and to identify changes in balance.

Membranous labyrinth

At about 22 days, a thickening of ectoderm on either side of the hindbrain develops; this is the **otic placode** (Figure 44.1). The placode invaginates forming a pit that later becomes separated from the ectoderm, forming the **otic vesicle** (or otocyst) deep to the ectoderm. The otic vesicle is surrounded by mesoderm that will

Embryology at a Glance, First Edition. Samuel Webster and Rhiannon de Wreede.

become the otic capsule, the cartilaginous precursor of the bony labyrinth.

The upper parts of the otic vesicle will develop into the **utricle, semicircular canals** and **endolymphatic duct** (the part of the inner ear involved in balance). The lower portion will become the **saccule,** which develops a tubular outgrowth that becomes the **cochlear duct** (the part of the inner ear involved in hearing).

Mesenchyme surrounding the cochlear duct becomes cartilaginous and two vacuoles appear within the cartilage. These vacuoles become the **scala tympani** and **scala vestibuli.** Positioned between them the cochlear duct (the **scala media**) remains, separated from the other spaces by membranes of mesenchyme (Figure 44.2). The spaces fill with perilymphatic fluid.

In the sixth week the cochlear duct grows into the surrounding mesenchyme, spiralling. It completes 2.5 spirals by the end of week 8 and is fully developed by the end of the second trimester (Figure 44.3).

Epithelial cells of the cochlear duct differentiate into the **sensory cells** and **tectorial membrane** of the auditory system, collectively called the **Organ of Corti**. These cells transmit auditory signals to the **vestibulocochlear nerve** (CN VIII). This space is also filled with endolymph that functions as a mechano-electric transducer for sound waves.

The foetus can hear external sounds by week 20.

The saccule and utricle remain connected to the cochlear duct via the **ductus reuniens** and the **utriculosaccular duct,** respectively. The saccule develops a group of sensory cells involved in translating vertical movements of the head.

Three flattened bud outgrowths protrude from the utricle to form three **semicircular canals**, each with an ampulla at one end, filled with endolymph. In each **ampulla** is a collection of sensory cells that aid balance. The utricle is involved in detecting horizontal movements of the head (Figures 44.3 and 44.4).

Bony labyrinth

The cartilage that surrounds the membranous labyrinth is ossified (weeks 16–24) and creates a perilymph-filled protective space for the inner ear. This area is connected to the subarachnoid space at the base of the brain through the **cochlear aqueduct** in the temporal bone. This is the petrous part of the temporal bone and is one of the hardest bones in the body. The vestibule of the inner ear contains the **oval window,** which is in contact with the **stapes** bone of the middle ear (Figures 44.4 and 44.5).

Middle ear

The middle ear consists of the **tympanic cavity,** the **pharyngotympanic tube** (or auditory or Eustachian tube) and the **ossicles** (Figures 44.4 and 44.5).

Endoderm of the first pharyngeal pouch (see Chapter 38) extends laterally and on contact with the ectoderm of the first pharyngeal cleft forms the **tubotympanic recess.** The distal part of the tubotympanic recess becomes the tympanic cavity and the proximal part becomes the pharyngotympanic tube (Figure 44.4).

Mesenchyme of the first and second pharyngeal arches develops into the ossicles. Specifically, the **malleus** and **incus** are derived from the first arch and the **stapes** from the second arch. These bones remain snug in surrounding tissue until the eighth month when the tissue dissipates and they become suspended within the developing cavity. As the tissue regresses the endoderm recedes but continues to line the cavity as a mesentery. The ligaments that will help to hold the ossicles in place develop from this mesentery.

Organisation of development of the muscles of the middle ear corresponds to the development of the bones. The **tensor tympani muscle** which inserts on to the malleus develops from the first pharyngeal arch and is innervated by the mandibular branch of the **trigeminal nerve.** The **stapedius** muscle which inserts on to the stapes develops from the second pharyngeal arch and is innervated by the **facial nerve** (see Chapters 38 and 39).

External ear

The external ear begins internally with the **tympanic membrane,** also known as the eardrum. This is the junction at which the internal endoderm-lined tympanic cavity meets the ectoderm of the external auditory meatus. Squeezed between these two layers is a layer of connective tissue (Figure 44.4).

Developing from the first pharyngeal cleft, the **external auditory meatus** retains cells in its proximal part which form a plug until the seventh month. This disintegrates leaving a layer of epithelia to form part of the tympanic membrane.

The **auricle** (or pinna) develops from six swellings or **hillocks** (hillocks of Hiss) formed by proliferating cells which can be seen from week 6. Hillocks 1–3 are from the first pharyngeal arch, and 4–6 are from the second arch. The developing auricle begins in a location caudal to the mandible, and with directional embryonic growth its position ascends to approximately the level of the eyes (Figure 44.6).

Clinical relevance

Many factors affecting the developing ear will result in **deafness**. Most are caused by genetic factors but some environmental factors are involved. The **rubella virus** can affect the development of the organ of Corti if infected in the seventh to eighth week of development. Other factors known to cause deafness are **cytomegalovirus, hyperbilirubinemia** (jaundice) and bacterial **meningitis.**

External ear defects are quite common as the fusion of the auricular hillocks is complicated. Anomalies are often associated with other malformations. Most common chromosomal disorders have ear malformation as one of their traits. For example, trisomy 13 (**Patau syndrome**) gives an underdeveloped tragus and lobule, trisomy 21 (**Down syndrome**) gives microtia, and **Ehlers–Danlos syndrome** causes lop ears which stand away from the head and are often larger than normal.

Less severe anomalies can include **pits** and **appendages** (or sinuses and tags). These are remnants of the developing hillocks.

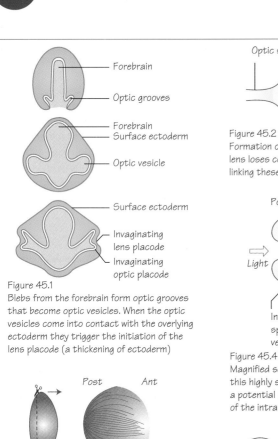

Figure 45.1
Blebs from the forebrain form optic grooves that become optic vesicles. When the optic vesicles come into contact with the overlying ectoderm they trigger the initiation of the lens placode (a thickening of ectoderm)

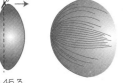

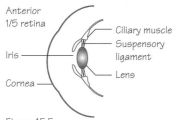

Figure 45.3
Formation of the primary lens fibres

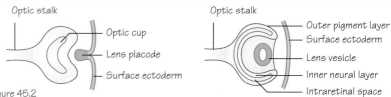

Figure 45.2
Formation of the lens vesicle and the optic cup. Note the outer and inner layers. The developing lens loses contact with the overlying ectoderm as it becomes the lens vesicle, and the stalk linking these structures with the forebrain becomes the optic stalk

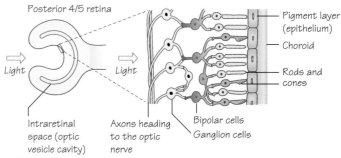

Figure 45.4
Magnified section of a part of the posterior 4/5 of the retina, showing all the cells that make up this highly specialised tissue. N.B. The intraretinal space becomes filled with cells, but there remains a potential space between the pigmented epithelia and the photoreceptive layer. This is a remnant of the intraretinal space and is termed the optic ventricle

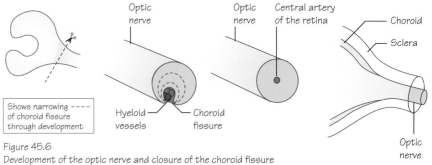

Figure 45.6
Development of the optic nerve and closure of the choroid fissure

Anterior 1/5 retina

Figure 45.5
Anterior 1/5 of the retina, and the iris and ciliary body that develop from it

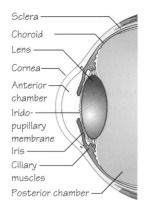

Figure 45.7
The iridopupillary membrane should regress but can persist after birth.
(a) This sagittal section shows the location of the iridopupillary membrane
(b) Anterior view of the iridopupillary membrane as it covers the pupil

Embryology at a Glance, First Edition. Samuel Webster and Rhiannon de Wreede.

Time period: weeks 3–10

Introduction
The development of the eye begins around day 22 with bilateral invaginations of the neuroectoderm of the forebrain (Figure 45.1).

Optic cup and lens
As the neural tube closes these invaginations become the **optic vesicles** and remain continuous with the developing third ventricle (Figure 45.1). Contact of these optic vesicles with the surface ectoderm induces the formation of the **lens placodes** (Figures 45.1 and 45.2).

As the optic vesicle invaginates it forms a double-walled structure, the **optic cup** (Figure 45.2). At the same time the lens placode invaginates and forms the **lens vesicle** which lies in the indent of the optic cup and is completely dissociated from the surface ectoderm.

Epithelial cells on the posterior wall of the lens vesicle lengthen anteriorly and become long fibres that grow forwards. It takes about 2 weeks for these fibres to reach the anterior cell wall of the vesicle. These are **primary lens fibres** (Figure 45.3). Secondary lens fibres form from epithelial cells located at the equator of the lens and are continuously added throughout life along the scaffold made by the primary fibres from the centre of the lens. These cells elongate and eventually lose their nuclei to become mature lens fibres. This occurs in early adulthood.

Retina
In the optic cup there is an outer layer that develops into the pigmented layer of the retina and an inner layer that becomes the neural layer.

The posterior four-fifths of the inner neural layer (Figure 45.4) consists of cells forming the rods and cones of the **photoreceptive layer**. Deep to this are the neurons and supporting cells. Deeper again lies a fibrous layer comprising the axons of these neurons, with axons leading towards the optic stalk that will develop into the **optic nerve**.

The anterior one-fifth of the inner layer remains one cell thick. It becomes parts of the **iris** and **ciliary body** (Figure 45.5). Ciliary muscle forms from mesenchyme that covers this part of the optic cup and is internally connected to the lens by the **suspensory ligament**. The complete iris forms from the inner retinal layer, the pigmented outer layer and a mesenchymal layer sandwiched between them, becoming the sphincter and dilator muscles.

Optic nerve
The optic cup remains attached to the forebrain via the **optic stalk** (Figure 45.2). Axons from the photoreceptor cells and other neurons within the retina run along the inner wall of the optic stalk.

Within the optic stalk is a groove, the **choroid fissure** (Figure 45.6), and within this lie the hyaloid blood vessels formed from invading mesenchyme. The number of neurons running through the optic stalk increases and during week 7 the choroid fissure closes forming a tunnel for the hyaloid artery which becomes the central artery of the retina (Figure 45.6). Continually increasing numbers of neurons fill the stalk and the lumen of the optic stalk is obliterated. By week 9 it is the **optic nerve**.

Meninges
The **choroid** and **sclera** of the eye are comparable to the pia mater and dura mater of the brain, respectively. They develop from the loose mesenchyme that surrounds the posterior part of the developing eye. The choroid is highly vascular and pigmented whereas the sclera is tougher and more fibrous. The sclera is continuous with the dura mater that surrounds the optic nerve (Figure 45.6).

Cornea
Loose mesenchyme surrounding the anterior part of the developing eye is split by vacuolisation, forming the anterior chamber. An inner iridopupillary membrane is created that degenerates, leaving open communication between the two fluid-filled spaces of the anterior and posterior chambers (Figure 45.7). The outer portion is continuous with the sclera and becomes the **cornea**.

The cornea has three parts: an epithelia layer from ectoderm, the mesenchyme part mentioned above and another epithelial layer that lines the anterior chamber. Neural crest cells contribute to the sclera and the cornea.

Extraocular muscles
Extraocular muscles include the inferior and superior oblique, medial, lateral, inferior and superior rectus and levator palpebrae superioris muscles. These muscles may develop from paraxial mesoderm of somitomeres 1–4 or from mesenchyme near the prechordal plate (a thickening of endoderm in the embryonic head) and are innervated by CN III, IV and VI.

Clinical relevance
Neonatal vision is tested in the 24 hours after birth and again at 6 weeks. Many abnormalities of the eye cause blindness, but not all. Most affect vision to some degree and as newborn infants' eyes are not aligned and an intermittent squint can develop, it is important that these tests are carried out.

Congenital cataracts can be caused by infection, such as rubella during pregnancy, or by hereditary factors, but often there is no known cause. Cataracts may be identified from a squint, difficulty in focusing or from the child holding his or her head at an odd angle to see, and in a few cases a clouding in the lens of the eye.

Congenital glaucoma is generally inherited and caused by an increase in intraocular pressure through excess fluid. It presents as watering of the eye with dilated pupils, irritability, and the cornea may be hazy. Treatments are aimed at decreasing the pressure through increasing the drainage of fluid from the eye or decreasing fluid production.

Coloboma is caused by incomplete closure of the choroid fissure. There are varying degrees, and usually only a cleft in the iris is left, which tends not to affect vision. The cleft can affect the eyelid, ciliary body, retina, choroid, lens or even the optic nerve, which would affect vision and can cause blindness in the most severe cases. There is currently no treatment.

Albinism affects pigmentation in the skin, hair, iris and retina, and is diagnosed from an eye examination. The lack of pigment in the iris and retina can lead to numerous eye conditions including macular hypoplasia, optic nerve hypoplasia, nystagmus, light sensitivity and overall poor vision.

Self-assessment MCQs

Embryonic and foetal periods

1 When does the embryonic period end?
(a) end of week 7
(b) beginning of week 8
(c) end of week 8
(d) end of week 9
(e) beginning of week 10

2 Which event in the female cycle is used to clinically date gestation?
(a) fertilisation
(b) first day of last menstruation
(c) last day of last menstruation
(d) length of menstruation
(e) ovulation

The first 18 days of development

3 Which hormone, secreted by the trophoblast, is important in maintaining the endometrium after implantation of the blastocyst?
(a) follicle stimulating hormone (FSH)
(b) human chorionic gonadotrophin
(c) luteinising hormone (LH)
(d) oestrogen
(e) testosterone

4 What is the final maturation process of spermatozoa that occurs in the female genital system?
(a) acrosome reaction
(b) capacitation
(c) cortical reaction
(d) spermatocytogenesis
(e) spermiogenesis

5 The acrosome and cortical reactions help prevent fertilisation of an ovum by more than one sperm. Which structure also helps prevent this?
(a) cortical granules
(b) cumulus cells
(c) fimbrae
(d) plasma membrane
(e) zona pellucida

6 Which structure is *not* formed from mesoderm?
(a) bone
(b) bone marrow
(c) dermis
(d) epidermis
(e) ureter

7 Which structure is *not* formed from endoderm?
(a) kidneys
(b) liver
(c) pancreas
(d) parathyroid glands
(e) tonsils

8 Which structure is *not* formed from ectoderm?
(a) epidermis
(b) central nervous system
(c) peripheral nervous system
(d) retina
(e) thymus gland

9 How many days after fertilisation does the blastocyst begin to implant into the uterine endometrium?
(a) 3 days
(b) 6 days
(c) 10 days
(d) 14 days
(e) 21 days

10 During which phase of the menstrual cycle does implantation occur?
(a) decidualization
(b) follicular
(c) menstrual
(d) proliferative
(e) secretory

11 During which phase of implantation does the inner cell mass rotate to become aligned with the decidua?
(a) adhesion
(b) apposition
(c) hatching
(d) implantation
(e) invasion

Gastrulation

12 Which signalling molecule helps cells of the epiblast move towards the primitive streak?
(a) brachyury
(b) fibroblast growth factor 4
(c) fibroblast growth factor 8
(d) transforming growth factor-beta
(e) vascular endothelial growth factor

13 Which process is *not* part of gastrulation?
(a) create cephalic–caudal axis
(b) create left–right axis
(c) create ventral–dorsal axis
(d) formation of the bilaminar disc
(e) formation of trilaminar disc

Neurulation

14 The caudal neuropore fails to close in an embryo. What condition will result if the foetus survives?
(a) cleft lip
(b) cerebral palsy

Embryology at a Glance, First Edition. Samuel Webster and Rhiannon de Wreede.
© 2012 John Wiley & Sons, Ltd. Published 2012 by John Wiley & Sons, Ltd.

(c) cystic fibrosis
(d) Down syndrome
(e) spina bifida

Body cavities

15 The diaphragm is formed from the septum transversum, pleuroperitoneal folds and some muscular ingrowth from the lateral body walls, and which other component?
(a) falciform ligament
(b) hepatic diverticulum
(c) lesser omentum
(d) oesophageal mesentery
(e) pleuropericardial folds

16 Which cavity links the thoracic and abdominal cavities in the fifth week of development?
(a) extra-embryonic cavity
(b) intra-embryonic cavity
(c) pericardioperitoneal canals
(d) peritoneal cavity
(e) pleural cavity

17 Which structure will form the adult pericardium?
(a) pleuropericardial folds
(b) pleuroperitoneal folds
(c) septum transversum
(d) somatic mesoderm
(e) splanchnic mesoderm

Musculoskeletal system

18 Which part of the somite will form the muscles of the anterior abdominal wall?
(a) dermatome
(b) dorsal myotome
(c) somitocoele
(d) sclerotome
(e) ventral myotome

19 What type of cell forms new bone, but is not completely surrounded by bone?
(a) chondrocyte
(b) hypertrophic chondrocyte
(c) osteoblast
(d) osteoclast
(e) osteocyte

20 Where does the secondary centre of ossification occur in a long bone such as the femur?
(a) articular cartilage
(b) diaphysis
(c) epiphysis
(d) growth plate
(e) periosteum

21 Which group of cells organises the development of structures in the cranial–caudal axis of the limb?
(a) apical ectodermal ridge
(b) hand plate

(c) proliferating zone
(d) ventral ectoderm
(e) zone of polarising activity

Cardiovascular system

22 What separates most of the single outflow tube of the early heart into the pulmonary trunk and aorta?
(a) atrioventricular septum
(b) conotruncal septum
(c) endocardial cushions
(d) septum primum
(e) septum secundum

23 What vessel allows blood to bypass the liver?
(a) ductus arteriosus
(b) ductus venosus
(c) sinus venosus
(d) umbilical artery
(e) umbilical vein

24 Which aortic arch artery forms part of the aorta?
(a) aortic arch I
(b) aortic arch II
(c) aortic arch III
(d) aortic arch IV
(e) aortic arch V

25 What structure grows within the heart to separate the atria from the ventricles?
(a) conotruncal septum
(b) endocardial cushion
(c) interventricular septum
(d) septum primum
(e) septum secundum

26 Two weeks after birth a cyanotic baby is found to be tachycardic, tachypnoeic and has a wide pulse pressure. What is the most likely diagnosis?
(a) mitral valve defect
(b) patent ductus arteriosus
(c) patent foramen ovale
(d) portosystemic shunt
(e) ventricular septal defect

Respiratory system

27 Which cell type forms a blood–air barrier with the endothelial cells of the pulmonary capillaries?
(a) ciliated epithelial cells
(b) columnar epithelial cells
(c) stratified epithelial cells
(d) type I alveolar cells
(e) type II alveolar cells

28 In the alveolar stage of lung development the number of alveoli increases, as does their volume, increasing the surface area available to gas exchange. When does the alveolar stage begin, approximately?
(a) 3 weeks
(b) 6 weeks

(c) 17 weeks
(d) 25 weeks
(e) 36 weeks

29 Because of its developmental origins, the trachea is most likely to form a fistula with what structure if it does not develop correctly?
(a) aorta
(b) diaphragm
(c) oesophagus
(d) pericardium
(e) spinal cord

Digestive system

30 What is the artery of the midgut?
(a) coeliac trunk
(b) ductus venosus
(c) inferior mesenteric artery
(d) superior mesenteric artery
(e) umbilical artery

31 By how much, and in which direction, does the midgut rotate?
(a) 90° counterclockwise
(b) 180° clockwise
(c) 270° counterclockwise
(d) 270° clockwise
(e) 360° counterclockwise

32 Which gut structure becomes separated into two parts by the urorectal septum?
(a) allantois
(b) cloaca
(c) hindgut
(d) perineum
(e) vitelline duct

33 Which organ develops from two separate dorsal and ventral buds of the gastrointestinal tract that rotate and fuse?
(a) gallbladder
(b) liver
(c) lung
(d) pancreas
(e) spleen

34 Daily doses of folic acid in the first 3 months of pregnancy is known to reduce the risk of spina bifida and which common gastrointestinal anomaly?
(a) cheiloschisis
(b) omphalocoele
(c) polyhydramnios
(d) tracheoesophageal fistula
(e) vitelline cyst

35 Which gastrointestinal anomaly is thought to be caused by failure of the intestines to return to the abdominal cavity after the normal umbilical herniation process?
(a) anal atresia
(b) omphalocoele

(c) polyhydramnios
(d) rectourethral fistula
(e) vitelline cyst

36 Identify one of the factors that can cause congenital hiatal hernia.
(a) gastric stenosis
(b) diaphragmatic hernia
(c) oesophageal atresia
(d) oesophageal stenosis
(e) shortened oesophagus

Urinary and reproductive systems

37 The tubes of the female reproductive system (i.e. the uterine tubes and the uterus) form from which embryonic structure?
(a) mesonephric ducts
(b) metanephros
(c) mesonephros
(d) paramesonephric ducts
(e) pronephros

38 What cells will differentiate into the transitional epithelium of the bladder?
(a) amniocytes
(b) ectoderm
(c) endoderm
(d) mesoderm
(e) yolk sac

39 Where do the definitive adult kidneys start to form?
(a) abdomen
(b) limb
(c) neck
(d) pelvis
(e) thorax

40 Where do the gonads start to form?
(a) abdomen
(b) limb
(c) neck
(d) pelvis
(e) thorax

41 Definitive nephrons (but not collecting ducts) develop from which embryonic structure?
(a) pronephros
(b) urachus
(c) ureteric bud
(d) urogenital sinus
(e) metanephric cap

Endocrine system

42 The inferior parathyroid glands are derived from cells from which embryonic structure?
(a) first pharyngeal arch
(b) second pharyngeal arch
(c) third pharyngeal arch

(d) fourth pharyngeal pouch
(e) sixth pharyngeal pouch

43 Which endocrine organ is derived from oral ectoderm and neural ectoderm?
(a) adrenal glands
(b) hypothalamus
(c) parathyroid glands
(d) pineal body
(e) pituitary gland

44 The medulla of the suprarenal (or adrenal) gland produces adrenaline and noradrenaline. What cell type is it formed from?
(a) dermamyotome
(b) epithelial
(c) mesoderm
(d) neural crest cell
(e) yolk sac

Head and neck

45 Which cranial nerve innervates many structures derived from the second pharyngeal arch?
(a) CN V (trigeminal nerve)
(b) CN VII (facial nerve)
(c) CN IX (glossopharyngeal nerve)
(d) CN X (vagus nerve)
(e) CN XII (hypoglossal nerve)

46 Which group of muscles are formed by the cells of the second pharyngeal arch?
(a) muscles of facial expression
(b) muscles of mastication
(c) muscles of the larynx
(d) muscles of the pharynx
(e) tensor tympani muscle

47 The mandible is a derivative of which pharyngeal arch?
(a) pharyngeal arch I
(b) pharyngeal arch II
(c) pharyngeal arch III
(d) pharyngeal arch IV
(e) pharyngeal arch VI

48 Calcitonin producing parafollicular cells of the thyroid gland are derived from cells of one of the pharyngeal pouches that migrate into the gland. Which pharyngeal pouch is the source of these cells?
(a) pharyngeal pouch I
(b) pharyngeal pouch II
(c) pharyngeal pouch III
(d) pharyngeal pouch IV
(e) pharyngeal pouch VI

49 The stapes bone is a derivative of which pharyngeal arch?
(a) pharyngeal arch I
(b) pharyngeal arch II
(c) pharyngeal arch III

(d) pharyngeal arch IV
(e) pharyngeal arch VI

Nervous system

50 From what cell layer does the central nervous system form?
(a) ectoderm
(b) endoderm
(c) mesoderm
(d) trophoblast
(e) yolk sac

51 In which week do the neuropores close?
(a) week 2
(b) week 3
(c) week 4
(d) week 5
(e) week 6

52 The dorsal root ganglion is a collection of sensory cell neurons that form peripheral nerves. What embryonic cell type differentiates to become these neurons?
(a) endoderm
(b) dermamyotome
(c) neural crest cell
(d) neuroepithelia
(e) paraxial mesoderm

53 The ventral horn of the spinal cord contains motor neurons. What part of the developing spinal cord becomes the ventral horn?
(a) alar plate
(b) basal plate
(c) marginal layer
(d) metencephalon
(e) pia mater

54 What is a function of the notochord?
(a) to initiate neurulation
(b) to initiate somite formation
(c) to organise dermatomes
(d) to regulate axon outgrowth
(e) to regulate limb development

The ear

55 In which week can the foetus hear external sounds?
(a) week 3
(b) week 6
(c) week 8
(d) week 10
(e) week 20

56 In the membranous labyrinth which structure is involved in translating vertical movements of the head?
(a) endolymphatic duct
(b) saccule
(c) scala tympani
(d) tectorial membrane
(e) utricle

57 The tubotympanic recess forms the tympanic cavity and part of which other structure?
 (a) external auditory meatus
 (b) pharyngotympanic tube
 (c) stapes
 (d) tensor tympani
 (e) tympanic membrane

58 In the seventh to eighth week the rubella virus can affect the development of which structure?
 (a) auricle
 (b) cochlea
 (c) organ of Corti
 (d) trigeminal nerve
 (e) tympanic membrane

The eye

59 Which neural cavity does the developing optic vesicle remain in contact with?
 (a) central canal
 (b) fourth ventricle
 (c) lateral ventricle (inferior horn)
 (d) lateral ventricle (main body)
 (e) third ventricle

60 Which structure degenerates to allow communication between the anterior and posterior chambers of the eye?
 (a) choroid fissure
 (b) iridopupillary membrane
 (c) lens vesicle
 (d) sclera
 (e) suspensory ligament

61 Which congenital abnormality of the eye can be caused by a rubella infection?
 (a) albinism
 (b) cataracts
 (c) glaucoma
 (d) macula hypoplasia
 (e) nystagmus

Self-assessment MCQ answers

Embryonic and foetal periods
1c, 2b

The first 18 days of development
3b, 4b, 5e, 6d, 7a, 8e, 9b, 10e, 11b

Gastrulation
12c, 13d

Neurulation
14e

Body cavities
15d, 16c, 17a

Musculoskeletal system
18e, 19c, 20c, 21e

Cardiovascular system
22b, 23b, 24d, 25d, 26b

Respiratory system
27d, 28e, 29c

Digestive system
30d, 31c, 32b, 33d, 34a, 35b, 36e

Urinary and reproductive systems
37d, 38c, 39d, 40a, 41e

Endocrine system
42d, 43e, 44d

Head and neck
45b, 46a, 47a, 48d, 49b

Nervous system
50a, 51c, 52c, 53b, 54a

The ear
55e, 56b, 57b, 58c

The eye
59e, 60b, 61b

Self-assessment EMQs

The first week of development

a. Blastocoele
b. Blastomere
c. Embryoblast
d. Epiblast
e. Exocoelomic membrane
f. Extra-embryonic mesoderm
g. Follicle
h. Hypoblast
i. Morula
j. Oogonia
k. Polar body
l. Trophoblast
m. Zona pellucida
n. Zygote

Choose the most appropriate option from the option list. Each option may be used once, more than once or not at all.

1 A spermatozoon fertilises an ovum to initially form what?
2 What is the name for the outermost cells of the blastocyst that will take part in the formation of the placenta?
3 Which layer must a spermatozoon penetrate to fertilise an ovum?
4 Which structure will develop the primitive streak, from which the germ layers will develop?
5 Cells of which structure produce human chorionic gonadotrophin?

Cardiovascular embryology

a. Coronary artery
b. Coronary sinus
c. Crista terminalis
d. Ductus arteriosus
e. Ductus venosus
f. Endocardial cushion
g. Ostium primum
h. Ostium secundum
i. Septum primum
j. Septum secundum
k. Sinus venosus
l. Truncus arteriosus
m. Umbilical artery
n. Vitelline artery

Choose the most appropriate option from the option list. Each option may be used once, more than once or not at all.

1 What structure develops to separate the primitive atrium from the primitive ventricle?
2 What structure allows blood to pass from the pulmonary trunk to the aorta?
3 What structure is formed by union of the major veins that drain into the primitive heart?
4 What is the outflow tract of the primitive heart called?
5 What structure forms the flap valve of the foramen ovale?

Gastrointestinal tract development

a. Appendix
b. Caecum
c. Colon
d. Duodenum
e. Gallbladder
f. Jejunum
g. Ileum
h. Liver
i. Oesophagus
j. Pancreas
k. Rectum
l. Stomach

Choose the most appropriate option from the option list. Each option may be used once, more than once or not at all.

1 The vitelline duct may persist as an outpocketing of which part of the gastrointestinal tract?
2 Severe vomiting in a newborn infant that contains bile may indicate a narrowing of the gastrointestinal tract where?
3 At which point along the adult gastrointestinal tract does the foregut end?
4 In which part of the gastrointestinal system may a fistula form that connects to the respiratory system?
5 What forms from the cloaca?

Neurulation and the central nervous system

a. Alar plate
b. Basal plate
c. Caudal neuropore
d. Cranial neuropore
e. Dorsal root ganglion
f. Mesencephalon
g. Metencephalon
h. Neural fold
i. Neural groove
j. Notochord
k. Primitive node
l. Primitive streak
m. Prosencephalon
n. Rhombencephalon

Choose the most appropriate option from the option list. Each option may be used once, more than once or not at all.

1 Where do neural crest cells come from?
2 Spina bifida occurs if what structure does not close properly?
3 From what does the hindbrain develop? (Choose the chronologically latest structure.)
4 Which structure signals to the midline ectoderm to form the neural tube?
5 Which structure is formed by neural crest cells?

Embryology at a Glance, First Edition. Samuel Webster and Rhiannon de Wreede.

Self-assessment EMQ answers

The first week of development

1n, 2l, 3m, 4d, 5l

Cardiovascular embryology

1f, 2d, 3k, 4l, 5j

Gastrointestinal tract development

1g, 2d, 3d, 4i, 5k

Neurulation and the central nervous system

1h, 2c, 3n, 4j, 5e

Glossary of medical conditions and terms

Accretion: Increase in size by gradual addition of smaller parts (e.g. a cell surrounding itself with matrix).

Achondroplastic dwarfism: Condition caused by limited long bone growth.

Acrosome: An organelle within the head of the sperm that carries enzymes.

Alar plates: Dorsal parts of the mantle layer of cell bodies that will become the dorsal sensory horns of the spinal cord.

Albinism: Defect in melanin production produces a lack of pigment.

Allantois: Extension of the primitive gut or yolk sac (depending on the stage of development) into the umbilical cord.

Amnioblasts: Cells of the amniotic membrane. From the epiblast.

Amnion: A thin, tough, membranous sac that encloses the embryo or foetus of a mammal, bird or reptile. It is filled with a serous fluid in which the embryo is suspended.

Anal atresia: No anus forms.

Anal membrane: Dorsal division of the cloacal membrane split by the urogenital septum.

Anencephaly: Failure of closure at the cranial end of the neural tube.

Anorectal canal: Posterior part of the cloaca when divided by the urogenital septum.

Antrum: A cavity formed between the layers of follicular cells.

Aortic arches: The vessels of the pharyngeal arches, link the heart and the dorsal aortae.

Aortic stenosis: A narrowing of the aorta.

Apical ectodermal ridge: Thickened ridge of ectoderm located along the distal edge of the limb bud.

Apoptosis: Programmed cell death (deliberate cell suicide).

Atrioventricular canal: Temporary connection between the primitive atria and ventricles.

Basal plates: Ventral parts of the mantle layer of cell bodies that will become the ventral motor horns of the spinal cord.

Bilaminar disc: Epiblast and hypoblast layers of the developing embryo.

Blastocoele: The fluid-filled central cavity of a blastocyst (or blastula).

Blastocyst: The early embryo as a sphere of cells with a fluid-filled central cavity (sometimes called the blastula).

Blastomeres: Any cell resulting from cleavage of a fertilised egg early in embryo development.

Bronchial buds: Lateral buds off the tracheal bud.

Buccopharyngeal membrane: Cranial area where endoderm is in direct contact with ectoderm, will form the mouth.

Bulboventricular sulcus: The junction between the bulbus cordis and the ventricle formed from the cardiac loop.

Bulbus cordis: Cranial bulge of the developing heart tube.

Capacitation: The process by which spermatozoa in the female genital tract become prepared for fertilisation.

Cardiac loop: Also called the bulboventricular loop, the first bend in the heart tube that occurs in week 4.

Cardinal veins: Two lateral vessels, the initial network of the veins that carry blood to the heart.

Cardiogenic field: Early developing cardiovascular tissue.

Cervical sinus: A sinus formed from pharyngeal clefts II, III and IV after the rapid growth of pharyngeal arch II forms the operculum.

Cheiloschisis: Cleft lip.

Chorionic: The outer membrane enclosing the embryo in reptiles, birds and mammals (contributes to the placenta in placental mammals).

Choroid fissure: A groove within the optic stalk where the hyaloid blood vessels are located.

Cleavage: A series of cell divisions in the ovum immediately following fertilisation.

Cloacal membrane: Caudal area where endoderm is in direct contact with ectoderm, will form the anus.

Coarctation of the aorta: Narrowing of the aorta.

Coelom: Fluid-filled body cavity lined by cells derived from mesoderm tissue in the embryo.

Coloboma: Incomplete closure of the choroid fissure.

Congenital adrenal hyperplasia: Autosomal recessive disease causing excessive steroid production.

Congenital cataracts: A clouding in the lens of the eye.

Congenital diaphragmatic hernia: Abdominal contents herniate into the thoracic cavity effecting development of the lungs.

Congenital glaucoma: Damage to the optic nerve through increased pressure in the eye.

Congenital hiatal hernia: A shortened oesophagus causes the stomach to be pulled into the thorax.

Congenital hypoparathyroidism: Incomplete development of the parathyroid glands resulting in low parathyroid hormones levels.

Congenital hypopituitarism: Decreased amounts of one or more of the hormones secreted by the pituitary gland.

Congenital hypothyroidism: Deficiency in thyroid hormone production leading to excessive sleeping and poor feeding.

Congenital subglottic stenosis: Abnormalities in the size and shape of the epiglottis can lead to breathing difficulties.

Conotruncal septum: A spiralled septum dividing the conus arteriosus and truncus arteriosus.

Conus arteriosus: Part of the outflow tract from the heart, will become the pulmonary trunk.

Conus cordis: Part of the heart tube that will become part of the left and right ventricles.

Corpus albicans: Scar tissue on the ovary formed from the corpus luteum.

Corpus luteum: A mature Graafian follicle that produces oestrogen and progesterone.

Cortical granules: Granules within the oocyte containing enzymes that bind the zona pellucida.

Cotyledons: Subunits of the placenta.

Craniosynostosis: Early closure of the cranial sutures.

Cryptorchidism: Undescended testes.

Cumulus oophorus: Also called the corona radiata, a layer of specialised cells that surround the oocyte.

Cytotrophoblast: The inner layer of the trophoblast.

Embryology at a Glance, First Edition. Samuel Webster and Rhiannon de Wreede.
© 2012 John Wiley & Sons, Ltd. Published 2012 by John Wiley & Sons, Ltd.

Decidualisation: Changes the endometrium undergoes in pregnancy.

Dermatomes: Regions of skin predominantly supplied by the sensory component of one spinal nerve.

Dermotome: Dorsal part of the somite that will develop into the dermis.

Dextrocardia: Heart lies on the right-hand side instead of the left.

Diencephalon: Part of the forebrain that will become the optic cup and stalk, pituitary gland, thalamus, hypothalamus and pineal body.

Differentiation: The process by which cells or tissues undergo a change toward a more specialised form or function.

Diploid: Cells with the full number of paired chromosomes (46 in humans).

Dorsal aortae: Two lateral vessels that are the beginning of the systemic blood system.

Dorsal mesentery: Develops as an outgrowth of the posterior body wall that passes over the gut and suspends it in the abdomen.

Dorsal roof plate: Dorsal bridging area for nerve fibres between the dorsal horns of the spinal cord.

Double inferior vena cavae: Caused by the persistence of the supracardinal veins.

Double superior vena cava: Caused by the persistence of the left anterior cardinal vein.

Ductus arteriosus: A vessel that acts as a shunt for blood passing between the pulmonary trunk and the aorta, allowing the majority of the blood to bypass the lungs.

Ductus venosus: A vessel that shunts blood from the umbilical vein to the inferior vena cava bypassing the liver.

Ectoderm: The outermost of the three primary germ layers of an embryo, from which the epidermis, nervous tissue and sense organs develop.

Ectrodactyly: Absence of a digit.

Embryo: The embryo of vertebrates is defined as the organism between the first division of the zygote (a fertilised ovum) until it becomes a foetus (8 weeks in humans).

Embryoblast: Any of the germinal disc cells of the inner cell mass in the blastocyst that form the embryo.

Embryology: The subdivision of developmental biology that studies embryos and their development.

Encephalocele: Herniation of the dura and neural tissue through a midline skull defect.

Endocardial cushions: Endocardium bulges that grow inwards and split the atrioventricular canal into two atrioventricular canals.

Endoderm: The innermost of the three primary germ layers of an animal embryo, developing into the gastrointestinal tract, the lungs and associated structures.

Endometrium: The glandular mucous membrane that lines the uterus.

Epiblast: The outer layer of cells in the inner cell mass that will form the embryo.

Epispadias: Urethral opening on the dorsal surface of the penis.

Exencephaly: Part of the brain exposed outside the skull.

Exstrophy: Ventral wall of the bladder is present outside of the abdominal wall.

External auditory meatus: The ear canal formed from the first pharyngeal cleft.

Extraembryonic: Of or being a structure that is outside the embryo.

Extraembryonic cavity: Fluid-filled cavity that forms surrounding the embryo, also known as the chorionic cavity.

Extra-uterine pregnancy: Also called ectopic pregnancy, implantation of the fertilised ovum somewhere other than the uterus.

Fertilisation: The process of a sperm fusing with an ovum, which eventually leads to the development of an embryo.

Fibrinoid deposits: Fibrin, placental secretions and dead trophoblast cells accumulate in the placenta.

Foetus: (alternatively, fetus or fœtus). An unborn vertebrate offspring after the embryo stage. In humans, a foetus develops from the end of week 8 of pregnancy (when the major structures have formed), until birth.

Follicle: Squamous epithelial cells that surround a primary oocyte.

Fontanelle: An area in the foetal skull where more than two bones meet.

Foramen cecum: The opening on the tongue of the thyroglossal duct.

Foramen ovale: An aperture between the right and left atria allowing blood to flow directly between atria.

Fossa ovalis: A depression left on the interior of the right atria after the foramen ovale closes.

Gap junctions: An intercellular network of protein channels that facilitates the cell–cell passage of ions, hormones and neurotransmitters.

Gastroschisis: Herniation of the bowel through the ventral abdominal wall.

Gastrulation: A phase early in the development of animal embryos, during which the morphology of the embryo is dramatically restructured by cell migration. In humans this process gives rise to the three embryonic germ layers.

Genital tubercle: Ventral fusion of the urogenital folds.

Goldenhar syndrome: Affects pharyngeal arches I and II and results in malformations of the facial bones and facial palsy.

Gonadal dysgenesis: Condition with male chromosomes but no testes.

Gondal ridge: Part of the urogenital ridge, contains cells that are the source of most of the genital system.

Graafian follicle: A mature follicle that expels the oocyte at ovulation.

Greater omentum: Develops from the dorsal mesentery.

Growth: In biology, growth is increase in size.

Gubernaculum: Part of the peritoneum attached to the gonad.

Haploid: Cells with half the number of paired chromosomes (23 in humans).

Homologous chromosomes: Two chromosomes that make up a 'pair' of chromosomes.

Hydrocephalus: Excess cerebrospinal fluid.

Hypertrophy: An increase in size brought about by an increase in cell size rather than division. It is most commonly seen in muscle that has been actively stimulated, the most well-known method being exercise.

Hypoblast: The inner layer of cells of the inner cell mass.

Hypophysial diverticulum: Also known as Rathke's pouch and outpocketing of oral ectoderm that will become the anterior lobe of the pituitary gland.

Hypospadias: Incomplete closure of the urethral folds.

Hyposplenism: No splenic function.

Implantation: The process by which a fertilised egg implants in the uterine lining.

Induction: The action of inducing cells to undergo change, usually in response to signalling molecules.

Intermediate horn: Small horn between dorsal and ventral horns of the spinal cord that contains neurons of the sympathetic nervous system between levels T1–L2.

Intermediate mesoderm: Mesodermal cells lateral to the paraxial mesoderm.

Interventricular foramen: A gap remaining between the muscular interventricular septum and the endocardial cushions.

Interventricular septum: A septum that splits the ventricles into left and right.

Intraembryonic cavity: Fluid-filled body cavity the forms within the embryo.

Intrauterine growth restriction: A condition where the placenta cannot supply the necessary nutrients to the foetus.

Labioscrotal swellings: Folds lateral to the urogenital folds.

Lacunae: An anatomical cavity, space or depression.

Laryngomalacia: Larynx can collapse during breathing.

Lateral plate mesoderm: Mesodermal cells lateral to the intermediate mesoderm.

Lens placodes: Areas of the surface ectoderm induced by contact with the optic vesicles.

Lens vesicle: Invagination of the lens placode.

Lesser omentum: Develops from the ventral mesentery.

Ligamentum arteriosum: The remnant of the the the ductus arteriosus.

Ligamentum venosus: The remnant of the ductus venosus.

Liver bud: Also called the hepatic diverticulum, a ventral bud from the distal foregut.

Mandibular process: Ventral division of the first pharyngeal arch.

Maxillary process: Dorsal division of the first pharyngeal arch.

Median umbilical ligaments: The remnant of the umbilical arties.

Meiosis: The process of cell division in sexually reproducing organisms that reduces the number of chromosomes in reproductive cells from diploid to haploid, leading to the production of gametes.

Membranous interventricular septum: A septum that grows inferiorly from the endocardial cushions to the muscular interventricular septum completing the interventricular septum.

Meromelia: Partial absence of a limb.

Mesencephalon: Part of the neural tube that will become the midbrain.

Mesoderm: The middle embryonic germ layer, lying between the ectoderm and the endoderm, from which connective tissue, muscle, bone and the urogenital and circulatory systems develop.

Mesonephric ducts: Also called Wolffian ducts, bilateral tubes that form from intermediate mesoderm, become an intermediate kidney and parts of the genital system.

Mesonephros: Second kidney to form.

Metanephric cap: Formed from intermediate mesoderm over the ureteric bud.

Metanephros: Third kidney to form, will become the adult kidney.

Metencephalon: Part of the hindbrain that will become the pons and cerebellum.

Mitosis: Division of a single cell into two 'daughter' cells, each with an identical number of chromosomes as the parent cell.

Morphogen: A substance that governs morphogenesis by emanating from a localised source to form a concentration gradient during embryonic development, metamorphosis or regeneration.

Morphogenesis: The processes that control the organised spatial distribution of cells that arise during embryonic development and give the shapes of tissues, organs and entire organisms.

Morula: The spherical embryonic mass of blastomeres formed before the blastula and resulting from cleavage of the fertilised ovum.

Muscular dystrophy: A group of diseases that causes muscular wasting.

Muscular interventricular septum: A septum that extends from the floor of the ventricles towards the endocardial cushions.

Myelencephalon: Part of the hindbrain that will become the medulla oblongata.

Myotome: The dorsal part of the somite that contains muscle precursor cells.

Nephrogenic cord: Part of the urogenital ridge, contains cells that are the source of most of the urinary system.

Neural crest cells: Cells derived from the ectoderm able to migrate extensively and generate many differentiated cell types (e.g. neurons, glial cells, the adrenaline-producing cells of the adrenal gland, pigmented cells of the epidermis).

Neural crests: Parts of neuroectoderm brought together that meet.

Neural folds: Sides of the neural groove.

Neural groove: Depression that forms along the neural plate.

Neural plate: A thick, flat bundle of ectoderm directly overlying the notochord that develops in the embryo into the neural tube and subsequently the nervous system.

Neural tube: See neural plate.

Neuroepithelia: Specialised cells that line the neurocoele.

Neurohypophysial diverticulum: Also known as the infundibulum, growth from the diencephalon that will become the posterior lobe of the pituitary gland.

Neuropores: Open ends of the neural tube.

Neurulation: A morphogenetic process in the embryonic development of the vertebrates, by which the neural plate folds into the neural tube.

Notochord: A rod of cells constituting the foundation of the axial skeleton, around which the segments of the vertebral column are formed.

Notochordal plate: Fusion of the notochordal process and the underlying endoderm.

Notochordal process: Midline extension of the primitive node.

Oligohydramnios: Reduction in the amount of amniotic fluid.

Omphalocoele: A herniation of the abdominal contents, covered by peritoneum and amnion, into the umbilicus.

Oocyte: A cell from which an egg or ovum develops by meiosis; a female gametocyte.

Oogonia: Diploid cells of the ovaries.

Operculum: A lid-like structure of mesenchyme formed from the rapid downward growth of pharyngeal arch II.

Optic cup: Double-walled invagination of the optic cup.

Optic stalk: The attachment that remains between the optic cup and the forebrain.

Optic vesicles: Bilateral invaginations of the neuroectoderm of the forebrain.

Ostium primum: A gap remaining between the septum primum and the endocardial cushions.

Ostium secundum: A gap remaining between the septum secundum and the endocardial cushions.

Otic placode: Bilateral thickening of ectoderm each side of the hindbrain.

Otic vesicle: Also called an otocyst, invagination of the otic placode that becomes isolated from the overlying ectoderm.

Ovum: The female reproductive cell or gamete of animals; egg.

Palatoschisis: Cleft palate.

Paramesonephric ducts: Also called Müllerian ducts, bilateral tubes that form from intermediate mesoderm lateral to the mesonephric ducts, become part of the genital system.

Paraxial mesoderm: Collection of mesodermal cells bilateral to the neural tube.

Patent ductus arteriosus: Failure of the ductus arteriosus to close.

Patent foramen ovale: Failure of the foramen ovale to close.

Pericardioperitoneal canal: Part of the body cavity that connect the thoracic cavity with the abdominal cavity.

Pharyngeal arch: Paired mesenchymal bar of cells and neural crest cells located in the ventrolateral parts of the head of the embryo.

Phocomelia: Proximal portion of a limb shortened.

Placenta praevia: Placenta develops in a low position in the uterus, covering part or all of the cervix.

Pleuroperitoneal folds: Membranes that separate the pleural and abdominal (peritoneal) cavity.

Pluripotent: Cells that can give rise to most, but not all, of the tissues necessary for foetal development.

Polar body: A minute cell produced and ultimately discarded in the development of an oocyte, containing little or no cytoplasm but having one of the nuclei derived from the first or second meiotic division.

Polydactyly: Presence of an extra digit.

Polyhydramnios: Excess amniotic fluid.

Portosystemic shunt: Failure of the ductus venosus to close.

Prechordal plate: Thickened part of the endoderm inferior to the buccopharyngeal membrane.

Primary follicle: Cuboidal epithelial cells that surround a primary oocyte.

Primary lens fibres: The first fibres to be laid down within the lens of the eye.

Primary oocyte: Oocyte during meiosis I.

Primary spermatocytes: Spermatogonia divide by meiosis and become secondary spermatocytes.

Primitive node: A round mound of cells at the cephalic end of the primitive streak.

Primitive pit: A circular depression in the centre of the primitive node.

Primitive streak: A structure involved in initiating gastrulation runs as a depression on the epiblastic surface of the bilaminar disc, is restricted to the caudal half of the embryo.

Primordial phallus: Structure formed after the genital tubercle elongates.

Proliferation: During cell reproduction one cell (the 'parental' cell) divides to produce daughter cells i.e. by mitosis. This is one form of tissue growth.

Pronephros: First transient kidney structure to form, redundant in humans.

Pronucleus: The haploid nucleus of a sperm or egg before fusion of the nuclei in fertilisation.

Prosencephalon: Part of the neural tube that will become the forebrain.

Proliferating zone: An area of cells under the apical ectodermal ridge where cells divide rapidly.

Rectourethral (urorectal) fistula: Failure of the urorectal septum to form completely, leaving contact between the rectum and urethra.

Rectovaginal fistula: Failure of the urorectal septum to form completely, leaving contact between the rectum and vagina.

Respiratory distress syndrome: Also known as hyaline membrane disease, results from a lack of surfactant produced in the lungs.

Respiratory diverticulum: Ventral bud from the proximal foregut.

Rhombencephalon: Part of the neural tube that will become the hindbrain.

Rhombomeres: Eight divisions of the hindbrain, inferior to the cephalic flexure.

Rickets: A condition resulting in weakened bones.

Robin sequence: Micrognathia with cleft palate, glossoptosis and absent gag reflex.

Round ligament of the liver: Or ligamentum teres hepatis.

Sacrococcygeal teratomas: Primitive streak cells are retained in the sacrococcygeal region and develop into tumours.

Sclerotome: Medial part of the somite, forms vertebrae.

Scoliosis: Lateral curvature of the spine.

Secondary follicle: Oocyte surounded by more than one layer of follicular epithelial cells.

Secondary spermatocytes: Haploid spermatocytes.

Septum primum: Membranous septum grows down from the roof of the atria separating them into left and right atria.

Septum secundum: Membranous septum grows down from the roof of the atria to the right of the septum primum.

Septum transversum: A sheet of mesodermal cells located between the pericardial cavity and the yolk sac stalk that forms a major part of the diaphragm, also involved in dividing the thoracic and abdominal cavities.

Shingles: Also called herpes zoster, viral disease that affects one dermatome.

Sinovaginal bulb: Outgrowth from the urogenital sinus.

Sinus venosus: Forms the inflow to the heart tube from the convergence of the major embryonic veins.

Sister chromatids: Identical copies of DNA.

Situs inversus: All internal organs are asymmetrical.

Somite: Paraxial mesodermal groups of cells forming either side of the midline, will form dermis, muscle and vertebrae.

Somitocoele: A lumen that develops in the centre of the somite.

Somitomeres: Paraxial mesodermal groups of cells forming either side of the midline, develop into somites.

Spermatids: Haploid cells formed from secondary spermatocytes dividing my meiosis II.

Spermatogenesis: Processes by which a spermatogonia becomes a spermatozoa.

Spermatogonia: Diploid germ cells that are stored in the testes, divide by meiosis to become primary spermatocytes.

Spermatozoa: Mature sperm ready for fertilisation.

Spermiogenesis: Formation of elongated spermatid, with tail and acrosome.

Spina bifida: Failure of the vertebrae is fuse completely.

Splanchnic mesoderm: The internal layer of the lateral plate mesoderm.

Stomodeum: A midline ectodermal depression ventral to the embryonic brain in the future face and surrounded by the mandibular arch.

Subcardinal veins: Longitudinal medial branching of the cardinal veins that form anastomoses with the posterior cardinal veins.

Sulcus limitans: A groove that divides the dorsal and ventral horns of the spinal cord.

Supracardinal veins: Longitudinal branching of the cardinal veins located between the cardinal nd subcardinal veins, form anastomoses with the posterior cardinal veins.

Syncytial knots: Grape-like nucleated clusters within the cytoplasm of the syncytiotrophoblast that occur late in the third trimester of pregnancy.

Syncytiotrophoblast: The syncytial outer layer of the trophoblast.

Syncytium: A mass of cytoplasm having many nuclei but no internal cell boundaries.

Syndetome: Located in the anterior and posterior edges of the somites between the cells of the myotome and sclerotome, contains tendon precursor cells.

Telecephalon: Part of the forebrain that will become the cerebral hemispheres.

Teratogen: Substance causing teratogenesis.

Teratogenesis: The formation of congenital malformations. Literally 'monster making' (Greek).

Tetralogy of Fallot: A collection of four congenital defects including pulmonary stenosis, an overriding aorta connecting both ventricles, a ventricular septal defect and hypertrophy of the right ventricle.

Thyroglossal duct: The connection between the thyroid gland and the tongue.

Totipotent: Cells able to divide and produce all the differentiated cells in an organism (stem cell).

Tracheoesophageal fistula: Abnormal connection between oesophagus and trachea.

Tracheoesophageal septum: Septum separating the respiratory bud and the oesophagus.

Treacher Collins syndrome: A genetic mutation causing the incomplete form of mandibulofacial dysostosis.

Trilaminar disc: Ectoderm, mesoderm and endoderm layers of the developing embryo.

Trophoblast: The outermost layer of cells of the blastocyst that attaches the fertilised ovum to the uterine wall and serves as a nutritive pathway for the embryo.

Truncus arteriosus: The cranial part of the primitive heart tube, becomes the ascending aorta and the pulmonary trunk.

Tubotympanic recess: The future pharyngotympanic tube (Eustachian/auditory) formed from the first pharyngeal pouch.

Ultimobranchial body: Collection of cells derived from the fifth pharyngeal pouch.

Umbilical veins: Two vessels that carry oxygenated blood from the placenta to the foetus.

Urachus: Forms from the allantois and becomes the median umbilical ligament.

Ureteric bud: Bud from the caudal end of the mesonephric duct.

Urethral plate: The structure created from a groove in the urogenital sinus lined with endodermal cells.

Urogenital folds: Also called urethral and clocal folds, mesenchymal folds surrounding the cloacal membrane.

Urogenital membrane: Ventral division of the cloacal membrane split by the urogenital septum.

Urogenital ridge: Forms from intermediate mesoderm.

Urogenital sinus: Anterior part of the cloaca when divided by the urogenital septum.

Urorectal septum: Mesodermal septum that divides the cloaca.

Uterovaginal primordium: Formed from paramesonephric ducts that fuse in the midline at pelvic level.

Vacuole: 1. A small cavity in the cytoplasm of a cell, bound by a single membrane and containing water, food or metabolic waste.
2. A small space or cavity in a tissue.

Vaginal plate: Fusion of the sinovaginal bulbs.

Ventral floor plate: Ventral bridging area for nerve fibres between the ventral horns of the spinal cord.

Ventral mesentery: Develops form the septum transversum.

Ventricular zone: Also called the ependymal layer, comprises thickening layers of neuroepithelia.

Vitelline cyst: Also called a omphalomesenteric duct cyst, cyst formed in the vitelline duct.

Vitelline stalk: Remaining contact between the gut tube and the yolk sac.

Vitelline vessels: Carry blood to and from the yolk sac.

Zona pellucida: The thick, solid, transparent outer membrane of a developed mammalian ovum.

Zone of polarising activity: A group of cells in the caudal mesenchyme of the limb bud.

Zygote: The cell formed by the union of two gametes, especially a fertilised ovum before cleavage.

Index